ISBN 0-8373-6630-5

DANTES-30

Rudman's Question and Answers on the

DANTES SUBJECT STANDARDIZED TESTS

Subject Examination In . . .

PHYSICAL SCIENCE

Questions and Answers

NLC

NATIONAL LEARNING CORPORATION

212 MICHAEL DRIVE, SYOSSET, NEW YORK 11791 (516) 921-8888

National Learning Corporation

212 Michael Drive, Syosset, New York 11791

(516) 921-8888
(800) 645-6337
FAX: (516) 921-8743
www.passbooks.com
sales @ passbooks.com
info @ passbooks.com

PRINTED IN THE UNITED STATES OF AMERICA

PASSBOOK®

NOTICE

This book is *SOLELY* intended for, is sold *ONLY* to, and its use is *RESTRICTED* to *individual*, bona fide applicants or candidates who qualify by virtue of having seriously filed applications for appropriate license, certificate, professional and/or promotional advancement, higher school matriculation, scholarship, or other legitimate requirements of educational and/or governmental authorities.

This book is *NOT* intended for use, class instruction, tutoring, training, duplication, copying, reprinting, excerption, or adaptation, etc., by:

(1) Other Publishers
(2) Proprietors and/or Instructors of "Coaching" and/or Preparatory Courses
(3) Personnel and/or Training Divisions of commercial, industrial, and governmental organizations
(4) Schools, colleges, or universities and/or their departments and staffs, including teachers and other personnel
(5) Testing Agencies or Bureaus
(6) Study groups which seek by the purchase of a single volume to copy and/or duplicate and/or adapt this material for use by the group as a whole without having purchased individual volumes for each of the members of the group
(7) Et al.

Such persons would be in violation of appropriate Federal and State statutes.

PROVISION OF LICENSING AGREEMENTS. — Recognized educational commercial, industrial, and governmental institutions and organizations, and others legitimately engaged in educational pursuits, including training, testing, and measurement activities, may address a request for a licensing agreement to the copyright owners, who will determine whether, and under what conditions, including fees and charges, the materials in this book may be used by them. In other words, a licensing facility *exists* for the legitimate use of the material in this book on other than an individual basis. However, it is asseverated and affirmed here that the materials in this book *CANNOT* be used without the receipt of the express permission of such a licensing agreement from the Publishers.

NATIONAL LEARNING CORPORATION
212 Michael Drive
Syosset, New York 11791

Inquiries re licensing agreements should be addressed to:
The President
National Learning Corporation
212 Michael Drive
Syosset, New York 11791

PASSBOOK SERIES®

THE *PASSBOOK SERIES®* has been created to prepare applicants and candidates for the ultimate academic battlefield – the examination room.

At some time in our lives, each and every one of us may be required to take an examination – for validation, matriculation, admission, qualification, registration, certification, or licensure.

Based on the assumption that every applicant or candidate has met the basic formal educational standards, has taken the required number of courses, and read the necessary texts, the *PASSBOOK SERIES®* furnishes the one special preparation which may assure passing with confidence, instead of failing with insecurity. Examination questions – together with answers – are furnished as the basic vehicle for study so that the mysteries of the examination and its compounding difficulties may be eliminated or diminished by a sure method.

This book is meant to help you pass your examination provided that you qualify and are serious in your objective.

The entire field is reviewed through the huge store of content information which is succinctly presented through a provocative and challenging approach – the question-and-answer method.

A climate of success is established by furnishing the correct answers at the end of each test.

You soon learn to recognize types of questions, forms of questions, and patterns of questioning. You may even begin to anticipate expected outcomes.

You perceive that many questions are repeated or adapted so that you can gain acute insights, which may enable you to score many sure points.

You learn how to confront new questions, or types of questions, and to attack them confidently and work out the correct answers.

You note objectives and emphases, and recognize pitfalls and dangers, so that you may make positive educational adjustments.

Moreover, you are kept fully informed in relation to new concepts, methods, practices, and directions in the field.

You discover that you are actually taking the examination all the time: you are preparing for the examination by "taking" an examination, not by reading extraneous and/or supererogatory textbooks.

In short, this PASSBOOK®, used directedly, should be an important factor in helping you to pass your test.

DANTES Subject Standardized Tests

INTRODUCTION

The DANTES (Defense Activity for Non-Traditional Education Support) subject standardized tests are comprehensive college and graduate level examinations given by the Armed Forces, colleges and graduate schools as end-of-subject course evaluation final examinations or to obtain college equivalency credits in the various subject areas tested.

The DANTES Examination Program enables students to obtain college credit for what they have learned on the job, through self-study, personal interest, correspondence courses or by any other means. It is used by colleges and universities to award college credit to students who demonstrate that they know as much as students completing an equivalent college course. It is a cost-efficient, time-saving way for students to use their knowledge to accomplish their educational goals.

Most schools accept the American Council on Education (ACE) recommendations for the minimum score required and the amount of credit awarded, but not all schools do. Be sure to check the policy regarding the score level required for credit and the number of credits to be awarded.

Not all tests are accepted by all institutions. Even when a test is accepted by an institution, it may not be acceptable for every program at that institution. Before considering testing, ascertain the acceptability of a specific test for a particular course.

Colleges and universities that administer DANTES tests may administer them to any applicant – or they may administer the tests only to students registered at their institution. Decisions about who will be allowed to test are made by the school. Students should contact the test center to determine current policies and schedules for DANTES testing.

Colleges and universities authorized to administer DANTES tests usually do so throughout the calendar year. Each school sets its own fee for test administration and establishes its own testing schedule. Contact the representative at the administering school directly to make arrangements for testing.

Checklist

For Students

- ✓ Visit **www.getcollegecredit.com** to obtain a list of tests, fact sheets, test preparation materials, participating colleges and universities, and much more.
- ✓ Contact your school advisor to confirm that the DSST you selected will fit into your curriculum.
- ✓ Consult the ***DSST Candidate Information Bulletin*** for answers to specific questions.
- ✓ Contact the test site to schedule your test.
- ✓ Prepare for your examination by using the fact sheet as a guide.
- ✓ Take the test.

If you would like a score report sent to your college or university, it is a good idea to bring the four-digit code with you. You must write the DSST Test Center Code for that institution on your answer sheet at the time of testing. DSST Test Center Codes are noted in the DSST Participating Colleges and Universities listing on the Web site.

If you prefer to send a score report to an institution at a later date, there is a transcript fee of $20 for each transcript ordered.

Thomson Prometric
DSST Program
2000 Lenox Drive, Third Floor
Lawrenceville, NJ 08648

Toll-free: 877-471-9860
609-895-5011

E-mail: pnj-dsst@thomson.com

MAKING A COLLEGE DEGREE WITHIN YOUR REACH

Today, there are many educational alternatives to the classroom—you can learn from your job, your reading, your independent study, and special interests you pursue. You may already have learned the subject matter covered by some college-level courses.

The DSST Program is a nationally recognized testing program that gives you the opportunity to receive college credit for learning acquired outside the traditional college classroom. Colleges and universities throughout the United States administer the program, developed by Thomson Prometric, year-round. Annually, over 90,000 DSSTs are administered to individuals who are interested in continuing their education. Take advantage of the DSST testing program; it speeds the educational process and provides the flexibility adults need, making earning a degree more feasible.

Since requirements differ from college to college, please check with the credit-awarding institution before taking a DSST. More than 1,800 colleges and universities currently award credit for DSSTs, and the number is growing every day. You can choose from 37 test titles in the areas of Social Science, Business, Mathematics, Applied Technology, Humanities, and Physical Science. A brief description of each examination is found on the pages that follow.

Reach Your Career Goals Through DSSTs

Use DSSTs to help you earn your degree, get a promotion, or simply demonstrate that you have college-level knowledge in subjects relevant to your work.

Save Time...

You don't have to sit through classes when you have previously acquired the knowledge or experience for most of what is being taught and can learn the rest yourself. You might be able to bypass introductory-level courses in subject areas you already know.

Save Money...

DSSTs save you money because the classes you bypass by earning credit through the DSST Program are classes you won't have to pay for on your way to earning your degree. You can use the money instead to take more advanced courses that can be more challenging and rewarding.

Improve Your Chances for Admission to College

Each college has its own admission policies; however, having passing scores for DSSTs on your transcript can provide strong evidence of how well you can perform at the college level.

Gain Confidence Performing at a College Level

Many adults returning to college find that lack of confidence is often the greatest hurdle to overcome. Passing a DSST demonstrates your ability to perform on a college level.

Make Up for Courses You May Have Missed

You may be ready to graduate from college and find that you are a few credits short of earning your degree. By using semester breaks, vacation time, or leisure time to study independently, you can prepare to take one or more DSSTs, fulfill your academic requirements, and graduate on time.

If You Cannot Attend Regularly Scheduled Classes...

If your lifestyle or responsibilities prevent you from attending regularly scheduled classes, you can earn your college degree from a college offering an external degree program. The DSST Program allows you to earn your degree by study and experience outside the traditional classroom.

Many colleges and universities offer external degree or distance learning programs. For additional information, contact the college you plan to attend or:

Center for Lifelong Learning
American Council on Education
One DuPont Circle NW, Suite 250
Washington, DC 20036
202-939-9475
www.acenet.edu
(Select "Center for Lifelong Learning" under "Programs & Services" for more information)

Fact Sheets

For each test, there is a Fact Sheet that outlines the topics covered by each test and includes a list of sample questions, a list of recommended references of books that would be useful for review, and the number of credits awarded for a passing score as recommended by the American Council on Education (ACE). *Please note that some schools require scores that are higher than the minimum ACE-recommended passing score.* It is suggested that you check with your college or university to determine what score they require in order to earn credit. You can obtain Fact Sheets by:

- Downloading them from www.getcollegecredit.com
- E-mailing a request to pnj-dsst@thomson.com
- Completing a Candidate Publications Order Form

DSST Online Practice Tests

DSST online practice tests contain items that reflect a *partial range of difficulty* identified in the Content Outline section on each Fact Sheet. There is an online DSST Practice Test in the following categories:

- Mathematics
- Social Science
- Business
- Physical Science
- Applied Technology
- Humanities

Although the online DSST Practice Test questions do not indicate the full range of difficulty you would find in an actual DSST test, they will help you assess your knowledge level. Each online DSST Practice Test can be purchased by visiting www.getcollegecredit.com and clicking on DSST Practice Exams.

TAKING DSST EXAMINATIONS

Earning College Credit for DSST Examinations

To find out if the college of your choice awards credit for passing DSST scores, contact the admissions office or counseling and testing office. The college can also provide information on the scores required for awarding credit, the number of credit hours awarded, and any courses that can be bypassed with satisfactory scores.

It is important that you contact the institution of your choice as early as possible since credit-awarding policies differ among colleges and universities.

Where to Take DSSTs

DSSTs are administered at colleges and universities nationwide. Each location determines the frequency and scheduling of test administrations. To obtain the most current list of participating DSST colleges and universities:

- Visit and download the information from www.getcollegecredit.com
- E-mail pnj-dsst@thomson.com

Scheduling Your Examination

Please be aware that some colleges and universities provide DSST testing services to enrolled students only. After you have selected a college or university that administers DSSTs, you will need to contact them to schedule your test date.

The fee to take a DSST is $60 per test. This fee entitles you to two score reports after the test is scored. One will be sent directly to you and the other will be sent to the college or university that you designate on your answer sheet. You may pay the test fee with a certified check or U.S. money order made payable to Thomson Prometric or you may charge the test fee to your Visa, MasterCard or American Express credit card. Note: The credit card statement will reflect a charge from Thomson Prometric for all DSST examinations. *(Declined credit card charges will be assessed an additional $25 processing fee.)*

In addition, the test site may also require a test administration fee for each examination, to be paid directly to the institution. Contact the test site to determine its administration fee and payment policy.

Other Testing Arrangements

If you are unable to find a participating DSST college or university in your area, you may want to contact the testing office of a local accredited college or university to determine whether a representative from that office will agree to administer the test(s) for you.

The school's representative should then contact the DSST Program at 866-794-3497 to arrange for this administration. If you are unable to locate a test site, contact Thomson Prometric for assistance at pnj-dsst@thomson.com or 866-794-3497.

Testing Accommodations for Students with Disabilities

Thomson Prometric is committed to serving test takers with disabilities by providing services and reasonable testing accommodations as set forth in the provisions of the *Americans with Disabilities Act* (ADA). If you have a disability, as prescribed by the ADA, and require special testing services or arrangements, please contact the test administrator at the test site. You will be asked to submit to the test administrator documentation of your disability and your request for special accommodations. The test administrator will then forward your documentation along with your request for testing accommodations to Thomson Prometric for approval.

Please submit your request as far in advance of your test date as possible so that the necessary accommodations can be made. Only test takers with documented disabilities are eligible for special accommodations.

On the Day of the Examination

It is important to review this information and to have the correct identification present on the day of the examination:

- Arrive on time as a courtesy to the test administrator.
- Bring a valid form of government-issued identification that includes a current photo and your signature (acceptable documents include a driver's license, passport, state-issued identification card or military identification). *Anyone who fails to present valid identification will not be allowed to test.*
- Bring several No. 2 (soft-lead) sharpened pencils with good erasers, a watch, and a black pen if you will be writing an essay.
- Do not bring books or papers.
- Do not bring an alarm watch that beeps, a telephone, or a phone beeper into the testing room.
- The use of nonprogrammable calculators, slide rules, scratch paper and/or other materials is permitted for some of the tests.

DSST SCORING POLICIES

Your DSST examination scores are reported only to you, unless you request that they be sent elsewhere. If you want your scores sent to your college, you must provide the correct DSST code number of the school on your answer sheet at the time you take the test. See the *DSST Directory of Colleges and Universities* on the Web site www.getcollegecredit.com.

If your institution is not listed, contact Thomson Prometric at 866-794-3497 to establish a code number. (Some schools may require a student to be enrolled prior to receiving a score report.)

Receiving Your Score Report

Allow approximately four weeks after testing to receive your score report.

Calling DSST Customer Service before the required four-week score processing time has elapsed will not expedite the processing of your scores. Due to privacy and security requirements, scores will not be reported to students over the telephone under any circumstance.

Scoring of Principles of Public Speaking Speeches

The speech portion of the *Principles of Public Speaking* examination will be sent to speech raters who are faculty members at accredited colleges that currently teach or have previously taught the course. Scores for the *Principles of Public Speaking* examination are available six to eight weeks from receipt by Thomson Prometric. If you take the *Principles of Public Speaking* examination and fail (either the objective, speech portion, or both), you must follow the retesting policy waiting period of six months (180 days) before retaking the entire exam.

Essays

The essays for *Ethics in America* and *Technical Writing* are optional and thus are not scored by raters. The essays are forwarded to the college or university that you designate, along with your score report, for their use in determining the award of credit. Before taking the *Ethics in America* or *Technical Writing* examinations, check with your college or university to determine whether the essay is required.

NOTE: *Principles of Public Speaking* speech topic cassette tapes and essays are kept on file at Thomson Prometric for one year from the date of administration.

How to Get Transcripts

There is a $20 fee for each transcript you request. Payment must be in the form of a certified check, U.S. money order payable to Thomson Prometric, or credit card. Personal checks and debit cards are NOT an acceptable method of payment. One transcript may include scores for one or more examinations taken. To request a transcript, download the Transcript Order Form from www.getcollegecredit.com.

DESCRIPTION OF THE DSST EXAMINATIONS

Mathematics

• **Fundamentals of College Algebra** covers mathematical concepts such as fundamental algebraic operations; linear, absolute value; quadratic equations, inequalities, radials, exponents and logarithms, factoring polynomials and graphing. The use of a nonprogrammable, handheld calculator is permitted.

• **Principles of Statistics** tests the understanding of the various topics of statistics, both qualitatively and quantitatively, and the ability to apply statistical methods to solve a variety of problems. The topics included in this test are descriptive statistics; correlation and regression; probability; chance models and sampling and tests of significance. The use of a nonprogrammable, handheld calculator is permitted.

Social Science

• **Art of the Western World** deals with the history of art during the following periods: classical; Romanesque and Gothic; early Renaissance; high Renaissance, Baroque; rococo; neoclassicism and romanticism; realism, impressionism and post-impressionism; early twentieth century; and post-World War II.

• **Western Europe Since 1945** tests the knowledge of basic facts and terms and the understanding of concepts and principles related to the areas of the historical background of the aftermath of the Second World War and rebuilding of Europe; national political systems; issues and policies in Western European societies; European institutions and processes; and Europe's relations with the rest of the world.

• **An Introduction to the Modern Middle East** emphasizes core knowledge (including geography, Judaism, Christianity, Islam, ethnicity); nineteenth-century European impact; twentieth-century Western influences; World Wars I and II; new nations; social and cultural changes (1900-1960) and the Middle East from 1960 to present.

• **Human/Cultural Geography** includes the Earth and basic facts (coordinate systems, maps, physiography, atmosphere, soils and vegetation, water); culture and environment, spatial processes (social processes, modern economic systems, settlement patterns, political geography); and regional geography.

• **Rise and Fall of the Soviet Union** covers Russia under the Old Regime; the Revolutionary Period; New Economic Policy; Pre-war Stalinism; The Second World War; Post-war Stalinism; The Khrushchev Years; The Brezhnev Era; and reform and collapse.

• **A History of the Vietnam War** covers the history of the roots of the Vietnam War; the First Vietnam War (1946-1954); pre-war developments (1954-1963); American involvement in the Vietnam War; Tet (1968); Vietnamizing the War (1968-1973); Cambodia and Laos; peace; legacies and lessons.

• **The Civil War and Reconstruction** covers the Civil War from presecession (1861) through Reconstruction. It includes causes of the war; secession; Fort Sumter; the war in the east and in the west; major battles; the political situation; assassination of Lincoln; end of the Confederacy; and Reconstruction.

• **Foundations of Education** includes topics such as contemporary issues in education; past and current influences on education (philosophies, democratic ideals, social/economic influences); and the interrelationships between contemporary issues and influences.

• **Life-span Developmental Psychology** covers models and theories; methods of study; ethical issues; biological development; perception, learning and memory; cognition and language; social, emotional, and personality development; social behaviors, family life cycle, extrafamilial settings; singlehood and cohabitation; occupational development and retirement; adjustment to life stresses; and bereavement and loss.

• **Drug and Alcohol Abuse** includes such topics as drug use in society; classification of drugs; pharmacological principles; alcohol (types, effects of, alcoholism); general principles and use of sedative hypnotics, narcotic analgesics, stimulants, and hallucinogens; other drugs (inhalants, steroids); and prevention/treatment.

• **General Anthropology** deals with anthropology as a discipline; theoretical perspectives; physical anthropology; archaeology; social organization; economic organization; political organization; religion; and modernization and application of anthropology.

• **Introduction to Law Enforcement** includes topics such as history and professional movement of law enforcement; overview of the U.S. criminal justice system; police systems in the U.S.; police organization, management, and issues; and U.S. law and precedents.

• **Criminal Justice** deals with criminal behavior (crime in the U.S., theories of crime, types of crime); the criminal justice system (historical origins, legal foundations, due process); police; the court system (history and organization, adult court system, juvenile court, pre-trial and post-trial processes); and corrections.

• **Fundamentals of Counseling** covers historical development (significant influences and people); counselor roles and functions; the counseling relationship; and theoretical approaches to counseling.

Business

• **Principles of Finance** deals with financial statements and planning; time value of money; working capital management; valuation and characteristics; capital budgeting; cost of capital; risk and return; and international financial management. The use of a nonprogrammable, handheld calculator is permitted.

• **Principles of Financial Accounting** includes topics such as general concepts and principles, accounting cycle and classification; transaction analysis; accruals and deferrals; cash and internal control; current accounts; long- and short-term liabilities; capital stock; and financial statements. The use of a nonprogrammable, handheld calculator is permitted.

• **Human Resource Management** covers general employment issues; job analysis; training and development; performance appraisals; compensation issues; security issues; personnel legislation and regulation; labor relations and current issues; an overview of the Human Resource Management Field; Human Resource Planning; Staffing; training and development; compensation issues; safety and health; employee rights and discipline; employment law; labor relations and current issues and trends.

• **Organizational Behavior** deals with the study of organizational behavior (scientific approaches, research designs, data collection methods); individual processes and characteristics; interpersonal and group processes and characteristics; organizational processes and characteristics; and change and development processes.

• **Principles of Supervision** deals with the roles and responsibilities of the supervisor; management functions (planning, organization and staffing, directing at the supervisory level); and other topics (legal issues, stress management, union environments, quality concerns).

• **Business Law II** covers topics such as sales of goods; debtor and creditor relations; business organizations; property; and commercial paper.

• **Introduction to Computing** includes topics such as history and technological generations; hardware/software; applications to information technology; program development; data management; communications and connectivity; and computing and society. The use of a nonprogrammable, handheld calculator is permitted.

• **Management Information Systems** covers systems theory, analysis and design of systems, hardware and software; database management; telecommunications; management of the MIS functional area and informational support.

• **Introduction to Business** deals with economic issues affecting business; international business; government and business; forms of business ownership; small business, entrepreneurship and franchise; management process; human resource management; production and operations; marketing management; financial management; risk management and insurance; and management and information systems.

• **Money and Banking** covers the role and kinds of money; commercial banks and other financial intermediaries; central banking and the Federal Reserve system; money and macroeconomics activity; monetary policy in the U.S.; and the international monetary system.

• **Personal Finance** includes topics such as financial goals and values; budgeting; credit and debt; major purchases; taxes; insurance; investments; and retirement and estate planning. The use of auxiliary materials, such as calculators and slide rules, is NOT permitted.

• **Business Mathematics** deals with basic operations with integers, fractions, and decimals; round numbers; ratios; averages; business graphs; simple interest; compound interest and annuities; net pay and deductions; discounts and markups; depreciation and net worth; corporate securities; distribution of ownership; and stock and asset turnover.

Physical Science

• **Astronomy** covers the history of astronomy, celestial mechanics; celestial systems; astronomical instruments; the solar system; nature and evolution; the galaxy; the universe; determining astronomical distances; and life in the universe.

• **Here's to Your Health** covers mental health and behavior; human development and relationships; substance abuse; fitness and nutrition; risk factors, disease, and disease prevention; and safety, consumer awareness, and environmental concerns.

• **Environment and Humanity** deals with topics such as ecological concepts (ecosystems, global ecology, food chains and webs); environmental impacts; environmental management and conservation; and political processes and the future.

• **Principles of Physical Science I** includes physics: Newton's Laws of Motion; energy and momentum; thermodynamics; wave and optics; electricity and magnetism; chemistry: properties of matter; atomic theory and structure; and chemical reactions.

• **Physical Geology** covers Earth materials; igneous, sedimentary, and metamorphic rocks; surface processes (weathering, groundwater, glaciers, oceanic systems, deserts and winds, hydrologic cycle); internal Earth processes; and applications (mineral and energy resources, environmental geology).

Applied Technology

• **Technical Writing** covers topics such as theory and practice of technical writing; purpose, content, and organizational patterns of common types of technical documents; elements of various technical reports; and technical editing. Students have the option to write a short essay on one of the technical topics provided. Thomson Prometric will not score the essay; however, for determining the award of credit, a copy of the essay will be forwarded to the college or university you've designated along with the score report or transcript.

Humanities

• **Ethics in America** deals with ethical traditions (Greek views, Biblical traditions, moral law, consequential ethics, feminist ethics); ethical analysis of issues arising in interpersonal and personal-societal relationships and in professional and occupational roles; and relationships between ethical traditions and the ethical analysis of situations. Students have the option to write an essay to analyze a morally problematic situation in terms of issues relevant to a decision and arguments for alternative positions. Thomson Prometric will not score the essay; however, for determining the award of credit, a copy of the essay will be forwarded to the college or university you've designated along with the score report or transcript.

• **Introduction to World Religions** covers topics such as dimensions and approaches to religion; primal religions; Hinduism; Buddhism; Confucianism; Taoism; Judaism; Christianity; and Islam.

• **Principles of Public Speaking** consists of two parts: Part One consists of multiple-choice questions covering considerations of Principles of Public Speaking; audience analysis; purposes of speeches; structure/organization; content/supporting materials; research; language and style; delivery; communication apprehension; listening and feedback; and criticism and evaluation. Part Two requires the student to record an impromptu persuasive speech that will be scored.

FREQUENTLY ASKED QUESTIONS ABOUT DSSTs

In order to pass the test, must I study from one of the recommended references?

The recommended references are a listing of books that were being used as textbooks in college courses of the same or similar title at the time the test was developed. Appropriate textbooks for study are not limited to those listed in the fact sheet. If you wish to obtain study resources to prepare for the examination, you may reference either the current edition of the listed titles or textbooks currently used at a local college or university for the same class title. It is recommended that you reference more than one textbook on the topics outlined in the fact sheet. You should begin by checking textbook content against the content outline included on the front page of the DSST fact sheet before selecting textbooks that cover the text content from which to study. Textbooks may be found at the campus bookstore of a local college or university offering a course on the subject.

Is there a penalty for guessing on the tests?

There is no penalty for guessing on DSSTs, so you should mark an answer for each question.

How much time will I have to complete the test?

Many DSSTs can be completed within 90 minutes; however, additional time can be allowed if necessary.

What should I do if I find a test question irregularity?

Continue testing and then report the irregularity to the test administrator after the test. This may be done by asking that the test administrator note the irregularity on the Supervisor's Irregularity Report or you can write to Thomson Prometric, DSST Program, 2000 Lenox Drive, Third Floor, Lawrenceville, NJ 08648, and indicate the form and question number(s) or circumstances as well as your name and address.

When will I receive my score report?

Allow approximately four weeks from the date of testing to receive your score report. Allow six to eight weeks to receive a score report for the *Principles of Public Speaking* examination.

Will my test scores be released without my permission?

Your test score will not be released to anyone other than the school you designate on your answer sheet unless you write to us and ask us to send a transcript elsewhere. Instructions about how to do this can be found on your score report. Your scores may be used for research purposes, but individual scores are never made public nor are individuals identified if research findings are made public.

If I do not achieve a passing score on the test, how long must I wait until I can take the test again?

If you do not receive a score on the test that will enable you to obtain credit for the course, you may take the test again after six months (180 days). Please do not attempt to take the test before six months (180 days) have passed because you will receive a score report marked <u>*invalid*</u> and your test fee will not be refunded.

Can my test scores be canceled?

The test administrator is required to report any irregularities to Thomson Prometric. The consequence of bringing unauthorized materials into the testing room, or giving or receiving help, will be the forfeiture of your test fee and the invalidation of test scores. The DSST Program reserves the right to cancel scores and not issue score reports in such situations.

What can I do if I feel that my test scores were not accurately reported?

Thomson Prometric recognizes the extreme importance of test results to candidates and has a multi-step quality-control procedure to help ensure that reported scores are accurate. If you have reason to believe that your score(s) were not accurately reported, you may request to have your answer sheet reviewed and hand scored.

The fees for this service are:

- $20 fee if requested within six months of the test date
- $30 fee if requested more than six months from the test date
- $30 fee if a re-evaluation of the *Principles of Public Speaking* speech is requested

The fee for this service can be paid by credit card or by certified check or U.S. money order payable to Thomson Prometric. Submit your request for score verification along with the appropriate fee or credit card information (credit card number and expiration date) to Thomson Prometric, DSST Program, 2000 Lenox Drive, Third Floor, Lawrenceville, NJ 08648. Include your full name, the test title, the date you took the test, and your Social Security number. Candidates will be notified if a scoring discrepancy is discovered within four weeks of receipt of the request.

What does ACE recommendation mean?

The ACE recommendation is the minimum passing score recommended by the American Council on Education for any given test. It is equivalent to the average score of students in the DSST norming sample who received a grade of C for the course. Some schools require a score higher than the ACE recommendation.

Who is NLC?

National Learning Corporation (NLC) has been successfully preparing candidates for 40 years for over 5,000 exams. NLC publishes Passbook® study guides to help candidates prepare for all DANTES and CLEP exams and almost every other type of exam from high school through adult career.

Go to our website — www.passbooks.com — or call (800) 632-8888 for information about ordering our Passbooks.

To get detailed information on the DSST program and DSST preparation materials, visit www.getcollegecredit.com.

If you are interested in taking the DSST exams, call 877-471-9860 or e-mail pnj-dsst@thomson.com.

dsst™

Fact Sheet

DANTES Subject Standardized Tests

PRINCIPLES OF PHYSICAL SCIENCE I

TEST INFORMATION

This test was developed to enable schools to award credit to students for knowledge equivalent to that which is learned by students taking the course. The school may choose to award college credit to the student based on the achievement of a passing score. The passing score for each examination is determined by the school based on recommendations from the American Council on Education (ACE). This minimum credit-awarding score is equal to the mean score of students in the norming sample who received a grade of C in the course. Some schools set their own standards for awarding credit and may require a higher score than the ACE recommendation. Students should obtain this information from the institution where they expect to receive credit.

CONTENT

The following topics, which are commonly taught in courses on this subject, are covered by this examination.

	Approximate Percent
I. Physics	60%
A. Newton's Laws of Motion -Motion -Force and acceleration -Inertia -Mass and weight -Action/reaction forces -Law of Circular Motion -Law of Universal Gravitation	
B. Energy and Momentum -Work -Power -Potential and kinetic energy -Momentum	
C. Thermodynamics -Temperature and heat -Heat transfer -Heat capacity -Changes in state -Mechanical equivalent of heat	
D. Waves and Optics -Reflection and refraction -Lenses and mirrors -Waves -Interference and diffraction	
E. Electricity and Magnetism -Electronic charges (Coulomb's Law) -Electric current -Potential differences -Ohm's Law -Magnetism -Electromagnetism -Circuits and applied electricity	
II. Chemistry	40%
A. Properties of Matter -Elements, compounds, mixtures, solutions -Physical and chemical changes and properties -Gas laws -Kinetic theory	

from the official announcement for instructional purposes

B. Atomic Theory and Atomic Structure
   -The Periodic Law
   -Periodicity of the elements
   -Atomic models of the atom
   -Ions and molecules
   -Stoichiometry
   -The nucleus and radioactivity

C. Chemical Reactions
   -Equations
   -Acids, bases, and salts
   -Catalysts

III. Miscellaneous
Some questions deal with general scientific principles or processes.

Questions on the tests require candidates to demonstrate the following abilities. Some questions may require more than one of the abilities.

- Knowledge of basic facts and terms (about 35-40% of the examination)
- Understanding of concepts and principles (about 35-40% of the examination)
- Ability to apply knowledge to specific problems and situations (about 25-30% of the examination)

SAMPLE QUESTIONS

1. Which of the following properties of fluids can serve to distinguish between a gas and a liquid?

   (A) Immiscibility
   (B) Malleability
   (C) Viscosity
   (D) Compressibility

2. A substance that alters the rate of a chemical reaction but is not itself altered is called a

   (A) catalyst
   (B) reactant
   (C) product
   (D) limiting agent

3. In which of the following cases is the vehicle described NOT accelerating?

   (A) A car traveling in a straight line, increasing its speed from 10 meters per second to 20 meters per second.
   (B) A bus traveling at a constant speed over the crest of a hill
   (C) A train traveling in a straight line at a constant speed
   (D) A car traveling at 15 meters per second around a curve

4. The best explanation for the location of the element helium in the periodic table is that it

   (A) has a filled outer shell
   (B) has a low density
   (C) is monatomic
   (D) is a nonmetal

5. Two forces, one of 300 newtons and the other of 400 newtons, act at right angles to each other. The magnitude of the resultant force is

   (A) 100 N
   (B) between 300 N and 400 N
   (C) 500 N
   (D) 700 N

6. Which of the following is true when the pendulum of a clock reaches the highest point of its arc?

   (A) The net force acting on the system is zero.
   (B) The kinetic energy is maximum
   (C) The potential energy is maximum
   (D) The frequency is zero

7. If the half-life of a certain isotope is one month, what portion of a sample of this isotope remains after two months?

   (A) None
   (B) One-fourth
   (C) One-third
   (D) Three-fourths

8. If the distance between a proton and an electron is doubled, the resulting attraction will be

   (A) four times as great
   (B) twice as great

(C) half as great
(D) one-fourth as great

9. The heat that is required to raise the temperature of 10 grams of a sample whose specific heat is 0.212 calories/gram °C from 30° C to 50° C would be

(A) 200.0 cal
(B) 42.4 cal
(C) 4.24 cal
(D) 2.0 cal

10. In the Earth-Moon system, if r is the distance between the two masses, the attracting force between them is

(A) directly proportional to r
(B) directly proportional to r^2
(C) inversely proportional to r
(D) inversely proportional to r^2

11. The work done in holding a 50-newton object at 2 meters above a table top is

(A) 980 J
(B) 100 J
(C) 25 J
(D) 0 J

12. Which of the following statements regarding the force on an object in circular motion is NOT true?

(A) It is inversely proportional to the period squared.
(B) It is inversely proportional to the mass.
(C) It is directly proportional to the velocity.
(D) It is directly proportional to the acceleration.

STUDYING FOR THE EXAMINATION

The following is a list of reference publications that were being used as textbooks in college courses of the same or similar title at the time the test was developed. Appropriate textbooks for study are not limited to those listed below. If you wish to obtain study resources to prepare for the examination, you may reference either the current edition of the following titles **or** textbooks currently used at a local college or university for the same class title. It is recommended that you reference **more than one textbook** on the topics outlined in this fact sheet. You should **begin by checking textbook content against the content outline** included on the front page of this Fact Sheet **before** selecting textbooks that cover the test content from which to study. Textbooks may be found at the campus bookstore of a local college or university offering a course on the subject.

Sources for study material suggested but not limited to the following:

Physics

Kirkpatrick, Larry D. and Gerald F. Wheeler. *Physics: A WorldView*. Fort Worth, TX: Saunders College Publishing, current edition.

Krauskopf, Konrad B. and Arthur Beisner. *The Physical Universe*. New York: McGraw-Hill, current edition.

Payne, Charles A., William R. Falls, and Charles J. Whidden. *Physical Science: Principles and Applications*. Dubuque, IA: William C. Brown Publishers, current edition.

Riban, David M. *Introduction to Physical Science*. New York: McGraw-Hill, current edition.

Sears, Francis W., Mark W. Zemansky, and Hugh D. Young. *College Physics*. Reading, MA: Addison-Wesley Publishing, current edition.

Chemistry

Brown, Theodore L., Eugene H. LeMay, Jr., and Bruce E. Burston. *Chemistry: The Central Science.* Englewood Cliffs, NJ: Prentice-Hall, Inc., current edition.

Holtzclaw, Jr., Henry F., William R. Robinson, and Jerome D. Odom. *College Chemistry With Quantitative Analysis.* Lexington, MA: D. C. Health, current edition.

Mortimer, Charles E. *Chemistry.* Belmont, CA: Wadsworth Publishing Co., current edition.

Current textbook used by a local college or university for a course on the subject.

CREDIT RECOMMENDATIONS

The Center for Adult Learning Educational Credentials of the American Council on Education (ACE) has reviewed and evaluated the DSST test development process and has made the following recommendations:

Area or Course Equivalent:	Physical Science I
Level:	Lower level baccalaureate
Amount of Credit:	Three (3) semester hours
Source:	ACE Commission on Educational Credit and Credentials

INFORMATION

Colleges and universities that would like additional information about the national norming, or assistance in local norming or score validation studies should write to: DSST Program, Mail Stop 11-P, The Chauncey Group International, 664 Rosedale Road, Princeton, New Jersey 08540.

It is advisable that schools develop a consistent policy about awarding credit based on scores from this test and that the policy be reviewed periodically. The Chauncey Group will be happy to help schools in this effort.

Correct Responses to sample questions: 1.D; 2.A; 3.C; 4.A; 5.C; 6.C; 7.B; 8.D; 9.B; 10.D; 11.D; 12.B.

I.N. D204311

Science and your career

Studying science helps us understand the discoveries that affect our daily lives. Every time we use a telephone, television, or computer, we are using a product of science. We use our knowledge of science when making decisions about our health and diet. Even common hobbies, such as cooking, gardening, and photography, rely on scientific principles.

By studying science, we learn how the universe works; we learn to observe, classify, measure, predict, interpret, and communicate data; and we develop the ability to think logically and solve problems. The skills and knowledge that come from studying science are important in many occupations.

There are almost 400,000 scientists employed in the United States, but 21 million workers use science on the job. For example, mechanics use scientific procedures when repairing or testing equipment. Physical therapists use biology and physics to rehabilitate patients. Journalists use scientific knowledge when writing about technology, health, or the environment. And scientific problem solving skills are necessary for most computer occupations.

Science courses are also important if you want an advanced education. College admissions officers often favor individuals who have taken science classes. Many colleges require at least 2 years of high school science courses, regardless of your intended major. If you want to be admitted into scientific and technical programs, you will probably need 3 or 4 years of high school science.

Advanced science

Agricultural scientists
Architects
Archivists and curators
Biological and medical scientists
Chemists
Chiropractors
Computer scientists, computer engineers, and systems analysts
Dentists
Engineering, science, and computer systems managers
Engineers
Forensic scientists
Foresters and conservation scientists
Geologists and geophysicists
Landscape architects
Meteorologists
Optometrists
Pharmacists
Physical therapists
Physician assistants
Physicians
Physicists and astronomers
Podiatrists
Respiratory therapists
Teachers, secondary and college (sciences)
Veterinarians

Careers for people interested in science

Although science skills are helpful in many occupations, some occupations rely heavily on science. If you have a strong interest in science, you might want to consider one of the following occupations.

Advanced science occupations require a thorough knowledge of scientific principles; a bachelor's degree with a number of college science courses is usually the minimum requirement. But many of these positions require a master's or doctoral degree.

Biologists. Biologists study living organisms and their relationship to each other and the environment. Most biologists specialize in one branch of biology—for example, microbiology, the study of microscopic organisms; zoology, the study of animals; or botany, the study of plants. These branches are then subdivided. For example, types of zoologists include mammalogists, who study mammals; ichthyologists, who study fish; ornithologists, who study birds; and herpetologists, who study reptiles and amphibians.

Chemists. Chemists search for new chemicals and find uses for existing ones. Their discoveries might be used to produce medicines or create stronger building materials. Some chemists specialize in one branch of chemistry. Biochemists, for example, study the chemical composition of living things. Physical chemists examine the physical characteristics of atoms, molecules, and chemical reactions.

Physicists. Physicists study the behavior of matter, the generation and transfer of energy, and the interaction of matter and energy. They study areas such as gravity, nuclear energy, electromagnetism, electricity, light, and heat. They might examine the structure of the atom, or design research equipment such as lasers. Physicists might also work in inspection, testing, or other production-related jobs.

Applied science occupations require workers to understand scientific principles and apply them to their work; some posthigh school science training is needed.

Agricultural scientists. Some types of scientists work to improve agriculture. Crop scientists study the genetic breeding and management of field crops. Soil scientists use soil physics, soil chemistry, and soil microbiology to enhance soil fertility and the growth of plants. Agronomists develop practical applications for discoveries in plant and soil science to produce high quality food.

Other scientists. There are many other branches of science. Geologists study the history and composition of our planet, including volcanoes and earthquakes. Oceanographers study the oceans and their movements. Meteorologists study the atmosphere, and some make weather predictions. Astronomers study the universe, trying to gain knowledge about the stars, planets, and galaxies.

Applied science

Aircraft mechanics, including engine specialists
Aircraft pilots
Broadcast technicians
Cardiovascular technologists and technicians
Clinical laboratory technologists and technicians
College and university faculty
Construction and building inspectors
Construction contractors and managers
Dental hygienists
Dental laboratory technicians
Dietitians and nutritionists
Dispensing opticians
Drafters
Electroneurodiagnostic technologists
Emergency medical technicians
Engineering technicians (all specialties)
Health information technicians
Health services managers
Licensed practical nurses
Nuclear medicine technologists
Occupational therapists
Occupational therapy assistants and aides
Photographers and camera operators
Physical therapists
Psychologists
Radiologic technologists
Recreational therapists
Registered nurses
Respiratory therapists
Science technicians
Electronic semiconductor processors
Speech-language pathologists and audiologists
Surgical technologists
Surveyors and mapping scientists

Practical application of science

Automotive body repairers
Automotive mechanics
Barbers and cosmetologists
Boilermakers
Chefs, cooks, and other kitchen workers
Dental assistants
Diesel mechanics
Electricians
Electronic equipment repairers
Elevator installers and repairers
Farm equipment mechanics
Farmers and farm managers
Firefighting occupations
Fishers, hunters, and trappers
Funeral directors
General maintenance mechanics
Heating, air-conditioning, and refrigeration technicians
Home appliance and power tool repairers
Industrial machinery repairers
Jewelers
Landscaping, groundskeeping, nursery, greenhouse, and lawn service occupations
Machinists and tool programmers
Medical assistants
Millwrights
Mobile heavy equipment mechanics
Motorcycle, boat, and small-engine mechanics
Nursing aides and psychiatric aides
Ophthalmic laboratory technicians
Pest controllers
Pharmacy technicians
Photographic process workers
Physical and corrective therapy assistants
Plumbers and pipefitters
Prepress workers
Printing press operators
Stationary engineers
Structural and reinforcing iron workers
Tool-and-die makers
Urban and regional planners
Vending machine servicers and repairers
Water and wastewater treatment plant operators
Water transportation occupations
Welders, cutters, and welding machine operators

Although many scientists specialize, most need to have knowledge in more than one branch of science. Agronomists, for example, combine their knowledge of biology, geology, chemistry, and mathematics to find better ways to grow food and conserve soil. They may also work closely with other scientists, such as microbiologists, biochemists, meteorologists, and entomologists.

Engineers. Engineers use the principles and theories of chemistry, physics, and mathematics to solve practical problems. They develop new products and improve systems and processes. Engineers design computers, generators, helicopters, spacecraft, and other devices. Engineering has many specialties. The largest are mechanical engineering, electrical and electronics engineering, and civil engineering.

Practical application occupations require familiarity with the basic principles of biology, chemistry, or physics; high school courses in these areas should be sufficient.

Mechanical engineers design and develop power-producing machines, such as internal combustion and rocket engines. Others design and develop power-using machines, such as refrigeration systems.

Electrical and electronics engineers design, develop, test, and supervise the production of electrical equipment. This includes computers, automobile ignition systems, and wiring and lighting in buildings. They also design communications, video, and radar equipment.

Civil engineers design and supervise the building of roads, bridges, tunnels, buildings, airports, harbors, and water supply, flood control, and sewage systems.

Technicians and technologists. Science and engineering technicians carry out the plans of scientists and engineers—setting up experiments, recording results, or testing product quality. They may also design simple experiments. These workers use testing and measuring devices and have a solid understanding of laboratory techniques.

Other technician occupations include drafters, who prepare technical drawings of structures and products; broadcast technicians, who install, repair, and operate radio and television equipment; and air-conditioning, refrigeration, and heating technicians.

Other careers that use science

Science skills are useful in many other occupations. For example, there are numerous occupations in health care, and all require knowledge of biology and other sciences. Physicians, nurses, dentists, veterinarians, and emergency medical technicians are just a few of the health occupations that require an understanding of science.

Many workers use chemistry and physics in their work. Chefs and cooks use chemistry when creating recipes and preparing food, because cooking ingredients are chemicals. Dietitians and nutritionists are also concerned with chemical content of foods. Farmers and horticulturists use fertilizers and pesticides, the products of chemistry. Electricians apply the principles of physics when wiring a building, and aircraft pilots use physics and meteorology to plot flight paths and fly planes.

Preparing for careers in science

Careers in science require orderly thinking, systematic work habits, and perseverance. If you are a student who is interested in scientific and technical careers, you should take as many science classes in high school as possible. Basic courses in earth science, biology, chemistry, and physics will form a solid foundation for further study. A strong background in mathematics is also important for those who want to pursue scientific, engineering, and technology-related careers.

The lists show occupations requiring different levels of scientific skill: Advanced, applied, or practical application. Advanced science occupations require a thorough knowledge of scientific principles; a bachelor's degree with a number of college science courses is usually the minimum requirement. Many of these positions require a master's or doctoral degree. Applied science occupations require workers to understand scientific principles and apply them to their work; some posthigh school science training is needed. Practical application occupations require familiarity with the basic principles of biology, chemistry, or physics; high school courses in these areas should be sufficient.

For more information on the level of education and training needed for specific occupations, consult the *Occupational Outlook Handbook*, available in most libraries, career centers, and placement offices and on the Internet at **http://www.stats.bls.gov/ocohome.htm**.

HOW TO TAKE A TEST

You have studied long, hard and conscientiously.

With your official admission card in hand, and your heart pounding, you have been admitted to the examination room.

You note that there are several hundred other applicants in the examination room waiting to take the same test.

They all appear to be equally well prepared.

You know that nothing but your best effort will suffice. The "moment of truth" is at hand: you now have to demonstrate objectively, in writing, your knowledge of content and your understanding of subject matter.

You are fighting the most important battle of your life—to pass and/or score high on an examination which will determine your career and provide the economic basis for your livelihood.

What extra, special things should you know and should you do in taking the examination?

BEFORE THE TEST

YOUR PHYSICAL CONDITION IS IMPORTANT

If you are not well, you can't do your best work on tests. If you are half asleep, you can't do your best either. Here are some tips:

1) Get about the same amount of sleep you usually get. Don't stay up all night before the test, either partying or worrying—DON'T DO IT!
2) If you wear glasses, be sure to wear them when you go to take the test. This goes for hearing aids, too.
3) If you have any physical problems that may keep you from doing your best, be sure to tell the person giving the test. If you are sick or in poor health, you really cannot do your best on any test. You can always come back and take the test some other time.

AT THE TEST

EXAMINATION TECHNIQUES

1) Read the general instructions carefully. These are usually printed on the first page of the exam booklet. As a rule, these instructions refer to the timing of the examination; the fact that you should not start work until the signal and must stop work at a signal, etc. If there are any *special* instructions, such as a choice of questions to be answered, make sure that you note this instruction carefully.

2) When you are ready to start work on the examination, that is as soon as the signal has been given, read the instructions to each question booklet, underline any key words or phrases, such as *least*, *best*, *outline*, *describe* and the like. In this way you will tend to answer as requested rather than discover on reviewing your paper that you *listed without describing*, that you selected the *worst* choice rather than the *best* choice, etc.

3) If the examination is of the objective or multiple-choice type – that is, each question will also give a series of possible answers: A, B, C or D, and you are called upon to select the best answer and write the letter next to that answer on your answer paper – it is advisable to start answering each question in turn. There may be anywhere from 50 to 100 such questions in the three or four hours allotted and you can see how much time would be taken if you read through all the questions before beginning to answer any. Furthermore, if you come across a question or group of questions which you know would be difficult to answer, it would undoubtedly affect your handling of all the other questions.

4) If the examination is of the essay type and contains but a few questions, it is a moot point as to whether you should read all the questions before starting to answer any one. Of course, if you are given a choice – say five out of seven and the like – then it is essential to read all the questions so you can eliminate the two which are most difficult. If, however, you are asked to answer all the questions, there may be danger in trying to answer the easiest one first because you may find that you will spend too much time on it. The best technique is to answer the first question, then proceed to the second, etc.

5) Time your answers. Before the exam begins, write down the time it started, then add the time allowed for the examination and write down the time it must be completed, then divide the time available somewhat as follows:
   - If 3-1/2 hours are allowed, that would be 210 minutes. If you have 80 objective-type questions, that would be an average of 2-1/2 minutes per question. Allow yourself no more than 2 minutes per question, or a total of 160 minutes, which will permit about 50 minutes to review.
   - If for the time allotment of 210 minutes there are 7 essay questions to answer, that would average about 30 minutes a question. Give yourself only 25 minutes per question so that you have about 35 minutes to review.

6) The most important instruction is to *read each question* and make sure you know what is wanted. The second most important instruction is to *time yourself properly* so that you answer every question. The third most important instruction is to *answer every question*. Guess if you have to but include something for each question. Remember that you will receive no credit for a blank and will probably receive some credit if you write something in answer to an essay question. If you guess a letter – say "B" for a multiple-choice question – you may have guessed right. If you leave a blank as an answer to a multiple-choice question, the examiners may respect your

feelings but it will not add a point to your score. Some exams may penalize you for wrong answers, so in such cases *only*, you may not want to guess unless you have some basis for your answer.

7) Suggestions
   a. Objective-type questions
      1. Examine the question booklet for proper sequence of pages and questions
      2. Read all instructions carefully
      3. Skip any question which seems too difficult; return to it after all other questions have been answered
      4. Apportion your time properly; do not spend too much time on any single question or group of questions
      5. Note and underline key words – *all, most, fewest, least, best, worst, same, opposite,* etc.
      6. Pay particular attention to negatives
      7. Note unusual option, e.g., unduly long, short, complex, different or similar in content to the body of the question
      8. Observe the use of "hedging" words – *probably, may, most likely,* etc.
      9. Make sure that your answer is put next to the same number as the question
      10. Do not second-guess unless you have good reason to believe the second answer is definitely more correct
      11. Cross out original answer if you decide another answer is more accurate; do not erase until you are ready to hand your paper in
      12. Answer all questions; guess unless instructed otherwise
      13. Leave time for review

   b. Essay questions
      1. Read each question carefully
      2. Determine exactly what is wanted. Underline key words or phrases.
      3. Decide on outline or paragraph answer
      4. Include many different points and elements unless asked to develop any one or two points or elements
      5. Show impartiality by giving pros and cons unless directed to select one side only
      6. Make and write down any assumptions you find necessary to answer the questions
      7. Watch your English, grammar, punctuation and choice of words
      8. Time your answers; don't crowd material

8) Answering the essay question

Most essay questions can be answered by framing the specific response around several key words or ideas. Here are a few such key words or ideas:

M's: manpower, materials, methods, money, management
P's: purpose, program, policy, plan, procedure, practice, problems, pitfalls, personnel, public relations

a. Six basic steps in handling problems:
    1. Preliminary plan and background development
    2. Collect information, data and facts
    3. Analyze and interpret information, data and facts
    4. Analyze and develop solutions as well as make recommendations
    5. Prepare report and sell recommendations
    6. Install recommendations and follow up effectiveness

b. Pitfalls to avoid
    1. *Taking things for granted* – A statement of the situation does not necessarily imply that each of the elements is necessarily true; for example, a complaint may be invalid and biased so that all that can be taken for granted is that a complaint has been registered
    2. *Considering only one side of a situation* – Wherever possible, indicate several alternatives and then point out the reasons you selected the best one
    3. *Failing to indicate follow up* – Whenever your answer indicates action on your part, make certain that you will take proper follow-up action to see how successful your recommendations, procedures or actions turn out to be
    4. *Taking too long in answering any single question* – Remember to time your answers properly

EXAMINATION SECTION

EXAMINATION SECTION

TEST 1

DIRECTIONS: Each question or incomplete statement is followed by several suggested answers or completions. Select the one that BEST answers the question or completes the statement. *PRINT THE LETTER OF THE CORRECT ANSWER IN THE SPACE AT THE RIGHT.*

1. Laws and theories are tested by 1.___
   A. opinions B. experiments
   C. superstitions D. observations

2. A standard of comparison for an experiment is 2.___
   A. control B. theory
   C. hypothesis D. observations

3. In one centimeter, there are 3.___
   A. 10 km B. 10 mm C. 100 mm D. 1000 mm

4. The ____ is the standard unit of volume in the metric system. 4.___
   A. liter B. gram C. ounce D. meter

5. The state of matter which has volume and definite shape is known as 5.___
   A. solid B. liquid C. gas D. plasma

6. There are ____ grams in a kilogram. 6.___
   A. .001 B. 10 C. 100 D. 1000

7. One meter is equal to ____ centimeters. 7.___
   A. 10 B. 100 C. 10000 D. 1000

8. Crowbars, seesaws, and scissors are 8.___
   A. inclined planes B. wedges
   C. screws D. levers

9. Al, Cu, N, and O are symbols for 9.___
   A. elements B. compounds C. mixtures D. gases

10. Metric units are based on powers of 10.___
    A. 1 B. 10 C. 100 D. 1000

11. The ____ is the standard unit of length in the metric system. 11.___
    A. liter B. gram C. meter D. foot

12. The state of matter which has a definite volume but NO definite shape is known as 12.___
    A. solid B. liquid C. gas D. plasma

13. The amount of space that matter occupies is its 13.___
A. mass B. weight C. volume D. density

14. The quantity MOST closely related to weight is 14.___
A. density B. mass C. force D. volume

15. Density is found by dividing the weight of an object by its 15.___
A. volume B. mass
C. surface area D. temperature

16. A balance or scale is used to measure 16.___
A. weight B. volume C. temperature D. expansion

17. Anything that occupies space and has weight is called 17.___
A. volume B. force C. weight D. matter

18. Ramps, pulleys, and levers make work 18.___
A. easier
B. harder
C. neither easier nor harder

Questions 19-22.

DIRECTIONS: Questions 19 through 22 refer to the diagram below.

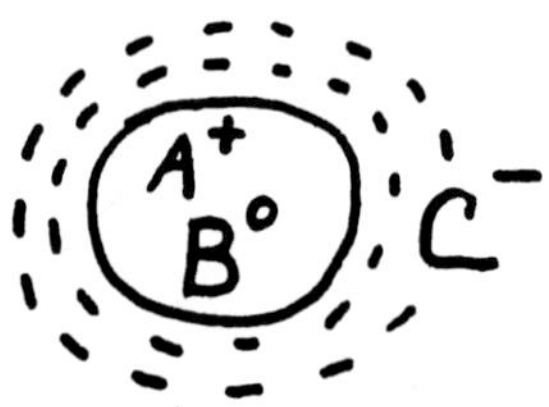

19. Particle (A^+) represents 19.___
A. electrons B. neutrons C. protons

20. Particle (B°) represents 20.___
A. neutrons B. protons C. electrons

21. Particle (C^-) represents 21.___
A. neutrons B. electrons C. protons

22. The above drawing or model illustrates a(n) 22.___
A. atom B. molecule of water
C. compound D. mixture

Questions 23-24.

DIRECTIONS: In answering Questions 23 and 24, refer to the following diagram. The diagram illustrates a model for the three states of matter, where (O) represents a molecule.

23. Which illustration represents a solid? 23. ____
A. A B. B C. C

24. Which illustration represents a gas? 24. ____
A. A B. B C. C

25. Sound does not travel in a 25. ____
A. liquid B. solid C. vacuum D. gas

26. If all the atoms in an object are the same, the object may be considered to be a(n) 26. ____
A. element
B. compound
C. mixture
D. both B and C

Questions 27-30.

DIRECTIONS: In Questions 27 through 30, identify the following changes as (A) CHEMICAL or (B) PHYSICAL.

27. Chopping wood 27. ____

28. Rusting of iron 28. ____

29. Boiling water 29. ____

30. $2H_2 + O_2 \rightarrow 2H_2O$ 30. ____

Questions 31-34.

DIRECTIONS: In answering Questions 31 through 34, classify the following types of energy in matter as (A) POTENTIAL or (B) KINETIC.

31. Wound up clock spring 31. ____

32. A weight raised above the floor 32. ____

33. Car rolling downhill 33. ____

34. Molecules vibrating due to heat 34. ____

Questions 35-38.

DIRECTIONS: In answering Questions 35 through 38, indicate whether the following reaction types are (A) EXOTHERMIC or (B) ENDOTHERMIC.

35. Burning match 35.____

36. Ice melting 36.____

37. Acid mixed with a metal 37.____

38. Boiling water 38.____

39. As particles of a gas slow down, they 39.____
A. condense B. evaporate C. melt D. expand

40. Which list is in order from small to large? 40.____
A. Electron, proton, atom, molecule, crystal
B. Proton, electron, atom, molecule, crystal
C. Neutron, electron, atom, molecule, crystal
D. Crystal, proton, atom, electron, molecule

41. The number of protons in an atom is equal to the number of 41.____
A. positrons
B. neutrons
C. electrons
D. all of the above

42. The arrangement of the atoms on the periodic table from Number 1 to Number 106 represents 42.____
A. atomic number
B. atomic weight
C. atomic mass
D. abundance of the atoms on earth

43. The LIGHTEST of all elements is 43.____
A. oxygen B. air C. hydrogen D. neon

44. ____ is(are) measured by volume. 44.____
A. Milk
B. Cottage cheese
C. Pretzels
D. Butter

45. The curved surface of a liquid in a container is a(n) 45.____
A. meniscus B. slope C. precipitate D. ohm

46. In an exothermic reaction, 46.____
A. boiling occurs
B. the thermometer breaks
C. heat is released
D. heat is absorbed

47. The BEST conductor of electricity is 47.____
A. aluminum B. sulfur C. oxygen D. copper

48. A(n) ____ is the charged particle that forms when an atom gains or loses an electron. 48.____
A. molecule B. nucleus C. ion D. muon

49. The atomic number represents the 49.___
A. number of protons B. number of electrons
C. number of neutrons D. atomic mass

50. An element having an atomic number 12 has 50.___
A. 10 protons B. 12 neutrons
C. both of the above D. none of the above

KEY (CORRECT ANSWERS)

1. B	11. C	21. B	31. A	41. C
2. A	12. B	22. A	32. A	42. A
3. B	13. C	23. A	33. B	43. C
4. A	14. A	24. C	34. B	44. A
5. A	15. A	25. C	35. A	45. A
6. D	16. A	26. A	36. B	46. C
7. B	17. D	27. B	37. B	47. D
8. D	18. A	28. A	38. A	48. C
9. A	19. C	29. B	39. A	49. A
10. B	20. A	30. A	40. A	50. D

TEST 2

DIRECTIONS: Each question or incomplete statement is followed by several suggested answers or completions. Select the one that BEST answers the question or completes the statement. *PRINT THE LETTER OF THE CORRECT ANSWER IN THE SPACE AT THE RIGHT.*

1. Work equals force times _____. (W=F× __) 1.____
   A. resistance B. distance C. ohms D. effort

Questions 2-7.

DIRECTIONS: In answering Questions 2 through 7, identify the following as (A) ELEMENT, (B) COMPOUND, or (C) MIXTURE.

2. Hydrogen (H) 2.____
3. Water (H_2O) 3.____
4. Diamond (C) 4.____
5. Coffee and sugar 5.____
6. Salt (NaCl) 6.____
7. Salt and sand 7.____

Questions 8-10.

DIRECTIONS: In answering Questions 8 through 10, refer to the illustration below.

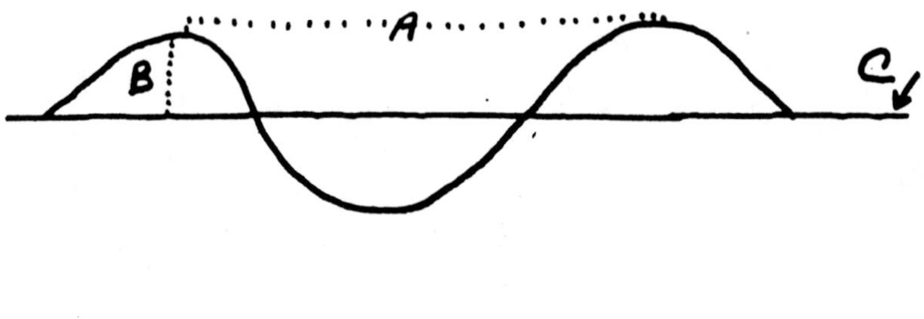

8. Wavelength is shown by letter 8.____
   A. A B. B C. C D. D
9. Amplitude is indicated by letter 9.____
   A. A B. B C. C D. D
10. Frequency of a wave means 10.____
    A. the size of a wave
    B. how often a wave occurs in a time period
    C. the speed of the wave

11. White light can be broken into the various colors by using a 11.____
    A. thermometer B. prism C. calorimeter D. mirror

12. A force can make things 12.____
    A. move B. stop
    C. change directions D. all of the above

13. As the temperature of air increases, the speed of sound 13.____
    A. decreases B. increases
    C. remains the same

14. If you want to increase the amount of sugar that dissolves in coffee, you should _____ the coffee 14.____
    A. heat B. cool C. evaporate

15. One may increase the _____ of a machine by reducing the friction. 15.____
    A. efficiency B. work input
    C. effort distance D. resistance distance

16. The bending of light is known as 16.____
    A. reflection B. resistance
    C. reverberation D. refraction

17. A lens that is thicker in the middle than at the ends is called 17.____
    A. conical B. concave C. convex D. chromatic

18. Sound is transmitted BEST through 18.____
    A. liquids B. gases C. solids D. a vacuum

19. The amplitude of a sound is determined by the 19.____
    A. wavelength B. pitch
    C. frequency D. none of the above

20. Of the following, which does NOT belong? 20.____
    A. Heat B. Temperature C. Light D. Sound

21. A(n) _____ solution is in equilibrium. 21.____
    A. unsaturated B. saturated
    C. supersaturated D. none of the above

22. Which law states that compounds are ALWAYS formed in the same mass ratio? 22.____
    A. Constant composition B. Conservation of mass
    C. Multiple proportions D. None of the above

23. If a stone is dropped from a tower and takes five seconds to reach the ground, its velocity will be _____ ft/sec. 23.____
    A. 432 B. 100 C. 160 D. none of these

24. Gas pressure depends upon 24.____
    A. the number of collisions with the container walls
    B. temperature
    C. the number of gas particles
    D. all of the above

25. When electrons are shared by two elements, a _____ bond is formed. 25.____
A. ionic B. covalent C. metallic D. halogen

26. Saying that an object looks cold is a(n) 26.____
A. inference
B. qualitative observation
C. objects property
D. quantitative observation

27. The farther a light source is from an object, the 27.____
A. more light reaches the object
B. greater the intensity
C. less light reaches the object
D. both A and B are correct

28. You could determine if a liquid is pure water or salt water WITHOUT tasting it by 28.____
A. testing the boiling point
B. testing the freezing point
C. evaporating the water to see if salt is present
D. all of the above

29. A 100 lb. boy sits on a seesaw three feet from the fulcrum. Where does a 75 lb. girl sit to balance the seesaw? 29.____
A. 1 ft. from the fulcrum
B. 2 ft. from the fulcrum
C. 4 ft. from the fulcrum
D. 4 ft. from the end of the seesaw

Questions 30-33.

DIRECTIONS: Column I lists four chemical equations, each of which is to be matched with one of the choices given in Column II. For each item of Column I, write in the space at the right the letter in front of the BEST choice in Column II.

COLUMN I	COLUMN II	
30. $Zn + 2HCl \rightarrow ZnCl_2 + H_2$	A. Synthesis reaction	30.____
31. $Na + Cl \rightarrow NaCl$	B. Decomposition	31.____
32. $NaI + PbNO_3 \rightarrow NaNO_3 + PbI$	C. Single replacement reaction	32.____
33. $2H_2O \rightarrow O_2 + 2H_2$	D. Double replacement reaction	33.____

Questions 34-35.

DIRECTIONS: Column I lists two items, each of which is to be matched with one of the choices given in Column II. For each item in Column I, write in the space at the right the letter in front of the BEST choice in Column II.

COLUMN I	COLUMN II	
34. Recoil of a bullet	A. Conservation of Momentum	34.___
35. Bowling ball striking pins	B. Law of Action and Reaction	35.___
	C. Law of Acceleration	

36. The symbol used to represent the gravitational constant for falling bodies is 36.___
    A. a B. v
    C. g D. none of the above

37. The average velocity can be determined from the formula 37.___
    A. $\frac{\text{time}}{\text{distance}}$ B. $\frac{\text{time}}{\text{acceleration}}$
    C. $\frac{\text{distance}}{\text{time}}$ D. distance x time

38. If a solution has a pH of 7, it is a(n) ____ solution. 38.___
    A. acid B. neutral C. base D. saturated

39. The time necessary for half of the nuclei in a radioactive isotope to decay is termed 39.___
    A. radioactivity B. half-life
    C. critical mass D. radiodating

40. If a solvent cannot dissolve any more solute, the solution is said to be 40.___
    A. concentrated B. saturated
    C. dilute D. suspended

41. According to Newton's First Law, every moving object has 41.___
    A. acceleration B. direction
    C. inertia D. none of the above

42. As the particles of matter are heated, they 42.___
    A. move slower
    B. remain unchanged
    C. move faster and closer together
    D. move faster and farther apart

43. Which of the objects below has the GREATEST momentum? 43.___
    A. $\frac{M = 10\text{ Kg}}{V=20\text{ m/sec}}$ B. $\frac{M = 40\text{ Kg}}{V=4\text{ m/sec}}$ C. $\frac{M = 5\text{ Kg}}{V=30\text{ m/sec}}$ D. $\frac{M = 200\text{ Kg}}{V=.5\text{ m/sec}}$

44. An object, attached to a string, is being swung in a circular path as shown at the right. If the string breaks when the object arrives at point *A*, in which direction will the object be traveling an instant later? 44. ____

A. ↑ B. ↓ C. ↖ D. →

45. The diagram at the right is of a ____ circuit. 45. ____

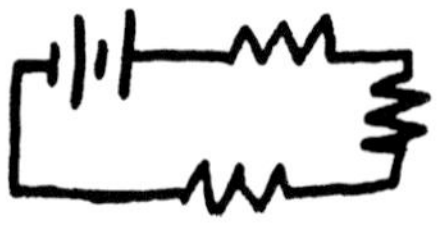

A. series
B. parallel
C. convex

Questions 46-48.

DIRECTIONS: In answering Questions 46 through 48, refer to the chart below.

SUBSTANCE	BOILING POINT °C	MELTING POINT °C	DENSITY g/cu c
A	78	-24	.78
B	56	18	1.32
C	26	-58	.52
D	125	52	2.65

46. If the temperature is 30°C in the room, which of the substances will be liquids? 46. ____
A. A and B
B. A, B, and C
C. B, C, and D
D. all of the above

47. If a student is given a sample of 200 cu cm of substance A, what will the mass of this sample be? 47. ____
A. 78 g B. 100 g C. 156 g D. 212 g

48. The LOWEST temperature at which all of the substances will be in the gas state is ____ °C. 48. ____
A. 100 B. 111 C. 125 D. 155

Questions 49-50.

DIRECTIONS: Match the diagrams in Column II with the correct description in Column I.

COLUMN I

49. Graph of freezing point 49. ____

50. Graph of boiling point 50. ____

COLUMN II

A. 

B. 

C. 

KEY (CORRECT ANSWERS)

1. B	11. B	21. D	31. A	41. C
2. A	12. D	22. B	32. D	42. D
3. B	13. B	23. C	33. B	43. A
4. A	14. A	24. D	34. B	44. D
5. C	15. A	25. B	35. A	45. A
6. B	16. D	26. B	36. C	46. A
7. C	17. C	27. C	37. C	47. C
8. A	18. C	28. D	38. B	48. C
9. B	19. B	29. C	39. B	49. A
10. B	20. D	30. A	40. B	50. C

EXAMINATION SECTION

TEST 1

DIRECTIONS: Each question or incomplete statement is followed by several suggested answers or completions. Select the one that BEST answers the question or completes the statement. *PRINT THE LETTER OF THE CORRECT ANSWER IN THE SPACE AT THE RIGHT.*

1. The metric unit of mass is the 1.___
   A. pound B. foot C. kilogram D. millimeter E. meter

2. The common multiplier in the metric system is 2.___
   A. 1 B. 2 C. 5 D. 10 E. 12

3. If the formula for acceleration = length/time^2 (/ means *divided by*), the cgs system units for acceleration MUST be 3.___
   A. m/s B. cm/s^3 C. m/s^2 D. mi/hr E. ft-lbs/in^2

4. How many grams in 0.25 Kg? 4.___
   A. 250 B. 0.00025 C. 400 D. 2500 E. 2.5

5. What part of a meter is 3mm? 5.___
   A. 3000 B. 0.003 C. 30,000 D. 0.03 E. 0.00003

6. The quantity obtained by dividing an object's mass by its volume is called the object's 6.___
   A. volume B. density C. mass D. weight E. solubility

7. ____ comes CLOSEST to expressing your height. 7.___
   A. 7m B. 70 cm C. 700cm D. 170cm E. 37.9cm

8. A cube of aluminum has a mass of 172.8 grams. If it has a volume of 64 cm^3, what MUST its density be? 8.___
   A. 10.8 g/cm B. 27 g/cm^3 C. 2.7 g/cm
   D. 2.7 g/cm^3 E. 108.8 g/cm^3

9. ____ is NOT a physical property. 9.___
   A. Formation of a compound B. Phase change
   C. Color D. Freezing
   E. Boiling

10. In order for a substance to burn, the element of ____ MUST be present. 10.___
    A. hydrogen B. nitrogen C. oxygen
    D. fire E. neon

11. A compound differs from a mixture in that 11.___
    A. each element keeps its identity
    B. substances may be combined in any ratio
    C. it may NOT have the same composition throughout
    D. it may NOT be easily separated
    E. all of the above

Questions 12-14.

DIRECTIONS: In Questions 12 to 14, you are to identify the materials indicated in the three important classes of matter as:
A. element
B. compound
C. mixture
D. none of these

12. Silver in jewelry 12.___

13. Bowl of oatmeal, sugar, and milk 13.___

14. Rust on a metal box (iron oxide) 14.___

15. When elements combine chemically, they form 15.___
A. mixtures B. reactions C. compounds
D. explosions E. allotropes

16. ____ is the CORRECT name for the compound indicated by the formula CuO. 16.___
A. Copper oxygen B. Copper oxide
C. Carbonate D. Carbon monoxide
E. Oxygen of copper

17. Which of these families is LEAST likely to form compounds? 17.___
A. Sulfur B. Nitrogen C. Noble gases
D. Oxygen E. Alkali metals

Questions 18-19.

DIRECTIONS: In Questions 18 and 19, you are to match the name of the element in Column I with its chemical symbol in Column II.

COLUMN I	COLUMN II	
18. Iron	A. I	18.___
19. Carbon	B. C	19.___
	C. Ca	
	D. Fe	

20. What is the TOTAL number of atoms present in each molecule of magnesium acetate, $Mg(C_2H_3O_2)_2$? 20.___
A. 16 B. 9 C. 10 D. 8 E. 15

Questions 21-24.

DIRECTIONS: Questions 21 through 24 refer to the equation below.

$$2\ KOH + H_2SO_4 \rightarrow K_2SO_4 \div 2\ H_2O$$

21. Which of the materials below would be called the reactants? 21.___
    A. KOH *only*
    B. H_2SO_4
    C. KSO_4 and KOH
    D. KOH and H_2SO_4
    E. none of the above

22. ____ atoms of oxygen are present in each molecule of K_2SO_4. 22.___
    A. 4
    B. 8
    C. 2
    D. 7
    E. 6

23. As the reaction is written above, is it CORRECTLY balanced? 23.___
    A. Yes, it is balanced.
    B. No, it is not balanced.
    C. It is impossible to tell if it is balanced.
    D. It is NOT a reaction so it cannot be balanced.
    E. Reactions are NEVER balanced.

24. How many molecules of water are formed in this reaction? 24.___
    A. Zero
    B. One
    C. Two
    D. Three
    E. Four

25. A(n) ____ reaction is indicated by a chemical change in which energy is released. 25.___
    A. hydrothermic
    B. isothermic
    C. endothermic
    D. thermic
    E. exothermic

26. What are the products of ALL neutralization reactions? 26.___
    A. Both are bases
    B. Both are acids
    C. Water and acid
    D. Water and a salt
    E. None of the above

27. Two liquids in two separate beakers stand on the pan of a balance which reads the mass of everything on it to be 284.6 grams. One of the beakers is picked up, its contents poured into the other beaker, and then set back down on the balance again. As the two liquids mix, a deep purple precipitate forms. 27.___
    What will be the NEW mass of the liquids and the beakers?
    A. More than 284.6 g
    B. 284.6 g
    C. Less than 284.6 g
    D. Impossible to predict with data given

28. The arrangement of the elements into the structure we call the Periodic Table is the result of the work of 28.___
    A. Dalton
    B. Rutherford
    C. Mendeleev
    D. Lavoisier
    E. Compton

29. The element immediately to the left of argon (one of the inert gases) on the periodic table is Chlorine. Thus, we know from Chlorine's position on the table that it has a 29.___
   A. strong tendency to gain an electron
   B. strong tendency to lose an electron
   C. weak tendency to gain an electron
   D. weak tendency to lose an electron
   E. tendency to neither gain nor lose electrons

Questions 30-32.

DIRECTIONS: Questions 30 through 32 refer to the subatomic particles:

A. Electrons
B. Protons
C. Neutrons
D. Electrons and protons
E. Protons and neutrons

30. Which particle or particles possess an electric charge? 30.___

31. Which particle is *directly* associated with the identity of the element? 31.___

32. Isotopes of an element all have the same number of ____. 32.___

33. When sulfur forms a compound, it USUALLY gains two electrons. What is the ion formed? 33.___
   A. S
   B. S^2
   C. S^{+2}
   D. S^{-2}
   E. None of the above

34. An element's ability to combine with other elements is shown by its 34.___
   A. mass number
   B. atomic number
   C. atomic weight
   D. valence number
   E. none of the above

35. The following compounds NaOH, KOH, $Mg(OH)_2$, $AL(OH)_3$, ALL have something in common. They are all 35.___
   A. elements B. salts C. insoluble D. acids E. bases

KEY (CORRECT ANSWERS)

1. C	11. D	21. D	31. E
2. D	12. A	22. A	32. B
3. C	13. C	23. A	33. D
4. A	14. B	24. C	34. D
5. B	15. C	25. E	35. E
6. B	16. B	26. D	
7. D	17. C	27. B	
8. D	18. D	28. C	
9. A	19. B	29. A	
10. C	20. E	30. D	

TEST 2

DIRECTIONS: Each question or incomplete statement is followed by several suggested answers or completions. Select the one that BEST answers the question or completes the statement. *PRINT THE LETTER OF THE CORRECT ANSWER IN THE SPACE AT THE RIGHT.*

1. How many neutrons are in each nucleus of the nuclide $^{14}_{6}C$? (This nuclide may also be written as $_{6}C^{14}$). 1.____
   A. 6 B. 8 C. 14
   D. 20 E. None of the above

2. Arsenic, phosphorous, sulfur, and nitrogen are classified as 2.____
   A. nonmetals B. organic compounds C. metals
   D. ions E. compounds

3. Carbon-containing compounds in complex arrangements are ALL classified as 3.____
   A. organic compounds B. basic compounds
   C. inorganic compounds D. carbonates
   E. neutralizing compounds

4. The classification given to elements having properties of both metals and nonmetals is 4.____
   A. alloys B. metalloids C. metal mix
   D. allotropes E. none of the above

5. In order to change the phase or state of a material from a liquid to a gas, we would have to 5.____
   A. ionize it B. remove heat C. add heat
   D. remove cold E. add cold

6. The liquid referred to as the universal solvent is 6.____
   A. alcohol B. benzene C. oil
   D. water E. none of the above

7. The amount of heat needed to warm up a 15 gram sample of water from 20°C to 30°C would be 7.____
   A. 15 calories B. 150 calories C. 10 calories
   D. 150 kilocalories E. 15 kilocalories

8. The heated air in a chimney rises up the chimney by the process of 8.____
   A. convection B. conduction C. radiation
   D. transduction E. none of the above

9. A sprinter travels an average of 10 m/s. If he could run this rate for 30 seconds, how far would he travel? 9.____
   A. 300 m B. 10 m C. 30 m D. 600 m E. 60 m

10. What would be the mechanical advantage of a wheel and axle with the following measurements: 50 cm wheel and 10 cm axle? 10.___
   A. M.A. = 60   B. M.A. = 5   C. M.A. = .20
   D. M.A. = 0   E. None of the above

11. A rocket accelerates at a rate of 500 meters per second every second. 11.___
If the rocket starts at rest, what will be the velocity of the rocket in 20 seconds?
   A. 1,000 m/s   B. 10,000 m/s   C. 5,000 m/s^2
   D. 20,000 m/s   E. 10,000 m/s^2

12. A sled changes speed from 3 meters to 15 meters in a four second period of time. 12.___
The average acceleration of the sled is
   A. 12 m/s^2 B. 12 m/s   C. 3 m/s^2
   D. 4.5 m/s^2 E. 3 m/s

13. The force which causes an object to accelerate in such a way that it has circular motion is called 13.___
   A. friction   B. centripetal force   C. inertial force
   D. applied force   E. none of the above

14. According to Newton's Third Law of Motion, the reason that a balloon flies around the room after you blow it up and release it is 14.___
   A. the sum of the forces is unequal
   B. there is an equal force in the opposite direction
   C. resistance is greater than the force
   D. gravity draws it to the center of the area
   E. molecules of air in the balloon are increasing in kinetic energy

Questions 15-17.

DIRECTIONS: In answering Questions 15 through 17, identify the type of energy as
   A. solar
   B. mechanical
   C. kinetic
   D. potential
   E. chemical

15. Any kind of energy stored for future use 15.___

16. A wound-up clock spring 16.___

17. Molecule vibrating due to the addition of heat 17.___

18. A force of 40 pounds is used to slide a storage chest of 200 pounds across a floor 8 feet. How much work has been done? 18.___
   A. 1600 ft-lb   B. 320 ft-lb   C. 5 ft-lb
   D. 25 ft-lb   E. 200 ft-lb

19. The reason that different keys struck on a piano produce different sounds is because of the difference in 19.___

A. tone B. pitch C. frequency
D. intensity E. quality

20. Of the two diagrams below, which shows the wave association with a sound of HIGHER pitch? 20.___

A. 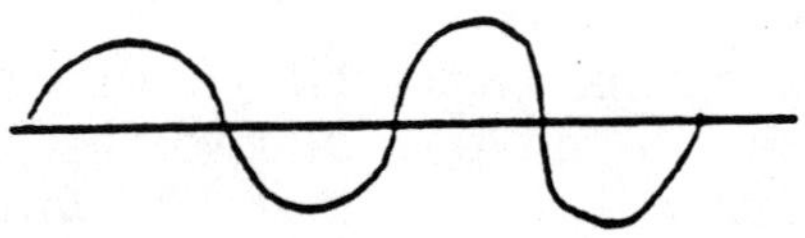

B. 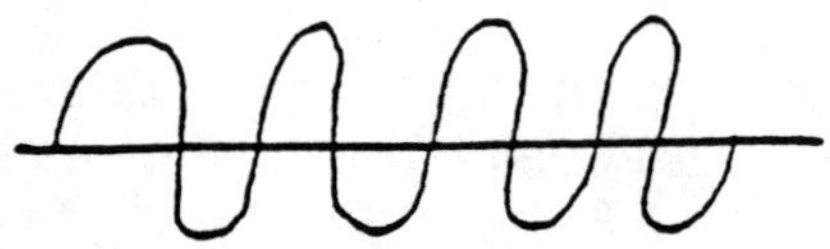

21. A blue shirt looks blue to you because it 21.___

A. reflects all colors but blue
B. absorbs all colors but blue
C. transmits all colors but red
D. reflects only red
E. absorbs only blue

22. Which of these lenses would have the SHORTEST focal length? 22.___

A. 

B. 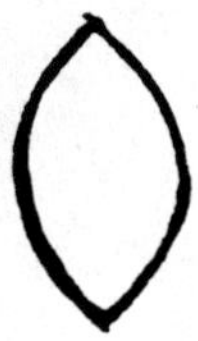

C. 

D. 

E. 

23. The bending of light is known as 23.___

A. reflection B. refraction C. absorption
D. diffraction E. reverberation

24. The unit in which the rate of charge flow is measured is 24.___

A. watts B. ohms C. volts D. amperes D. joules

25. A circuit with a 40 ohm resistance is connected to a 20-volt battery. 25.___
The current in the circuit is

A. 5 ohms B. 2 amps C. 0.5 amps D. 800 ohms E. 60 amps

Questions 26-27.

DIRECTIONS: Questions 26 and 27 refer to the following sketches of waves.

A. 

B. 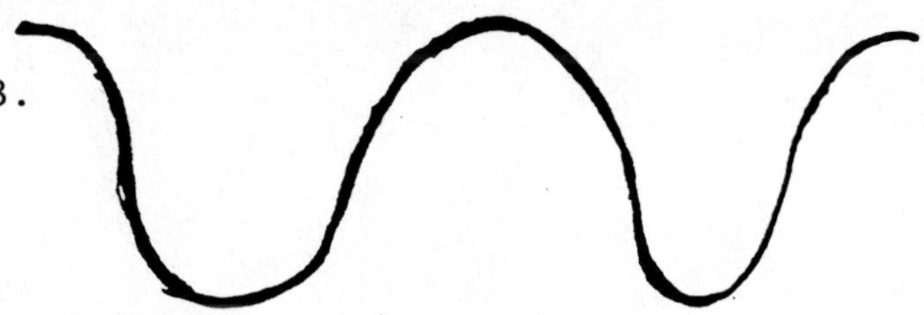

26. Which sketch shows a transverse wave? 26.___

27. Which sketch shows a longitudinal wave? 27.___

28. The individual MOST directly responsible for formulating laws describing the motion of objects in the universe was 28.___
   A. Einstein B. Galileo C. Alchemy
   D. Newton E. Aristotle

29. Pushing a refrigerator up a ramp into the back of a truck illustrates the use of which simple machine? 29.___
   A. Wedge B. Lever C. Inclined plane
   D. Pulleys E. Screw

30. If we wrap several turns of wire around a nail, place the nail near a compass, and connect the wire to a battery, the compass needle moves. 30.___
This occurs because of
   A. electromagnetism B. rectifiers
   C. voltmeter D. magnet
   E. thermal conduction

KEY (CORRECT ANSWERS)

1. B	11. B	21. B
2. A	12. C	22. A
3. A	13. B	23. B
4. B	14. A	24. C
5. C	15. D	25. C
6. D	16. B	26. B
7. C	17. C	27. A
8. A	18. B	28. D
9. A	19. C	29. C
10. B	20. B	30. A

EXAMINATION SECTION

TEST 1

DIRECTIONS: Each question or incomplete statement is followed by several suggested answers or completions. Select the one that BEST answers the question or completes the statement. *PRINT THE LETTER OF THE CORRECT ANSWER IN THE SPACE AT THE RIGHT.*

1. The metric system is based on multiples of 1.____
   A. 10 B. 100 C. 1,000 D. 10,000

2. A liter is slightly larger than a(n) 2.____
   A. ounce B. pint C. quart D. gallon

3. A standard for comparison in an experiment is the 3.____
   A. hypothesis B. guide
   C. theory D. control

4. Two thermometers appear below. 4.____
   Which thermometer is marked in celsius degrees?

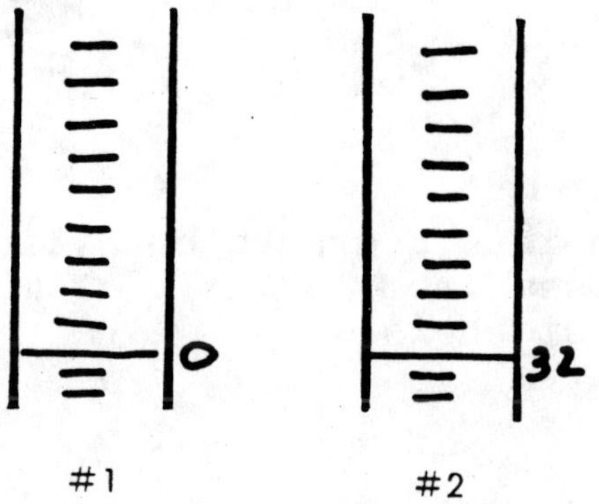

   A. #1 B. #2
   C. Neither one D. Both of them

5. _____ is CLOSEST to a gram of mass. 5.____
   A. One quart of milk
   B. Two paper clips
   C. Your physical science book
   D. A gram is too small to use as a unit of measurement

6. The figure below represents a portion of a meter stick. 6.____
   The arrow is pointing at

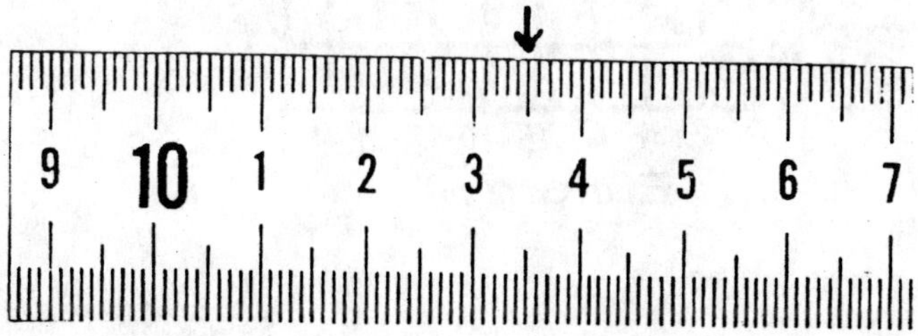

   A. 34 meters B. 13.4 cm C. 3.4 cm D. 13.4 mm

7. Which of the pieces of labware below represents a beaker. 7.___

A.

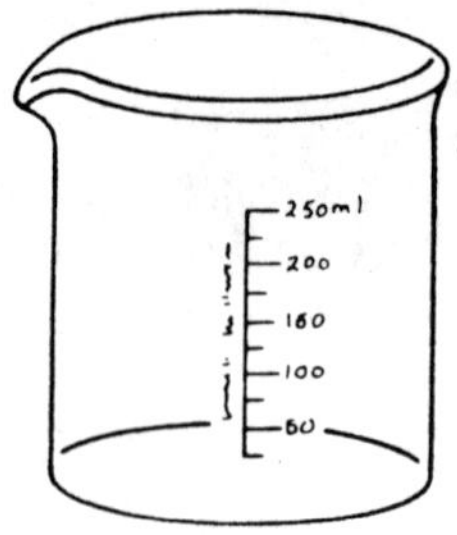

B.

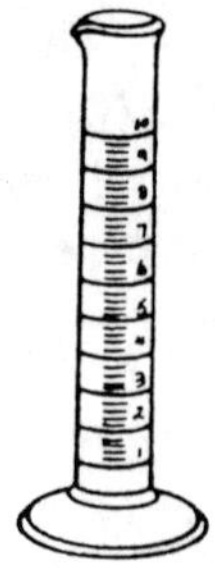

C.

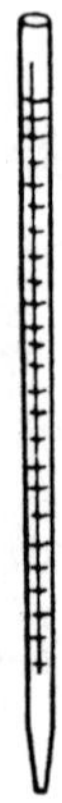

D.

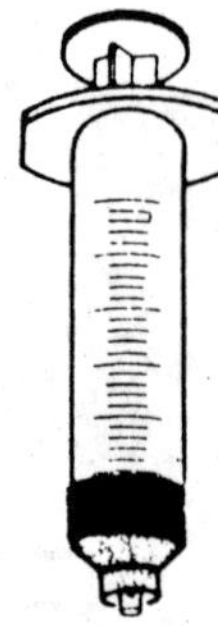

8. A stone is dropped into a graduated cylinder. The volume of water rises from 18 ml to 23 ml. 8.___
The volume of the stone is
  A. 0.0 ml
  B. 41.0 ml
  C. 5.0 ml
  D. the volume increases gradually

9. The temperature time graph below represents data collected as water is raised to the boiling point. 9.___
How hot was the water after nine minutes?

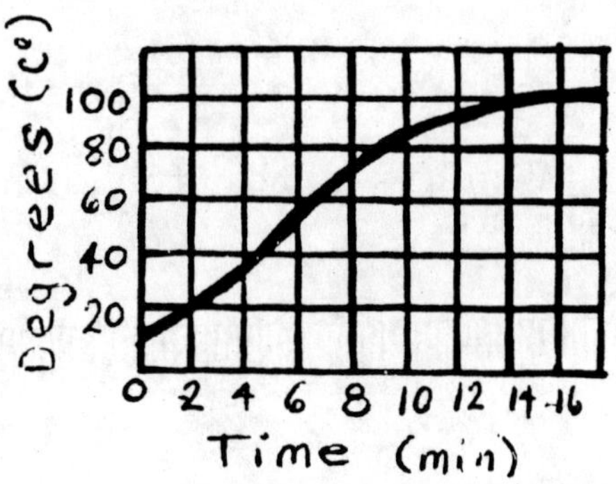

A. 80°C
B. 11°C
C. 20°C
D. Boiling hot

10. Only a(n) ____ force will produce motion. 10.___
   A. frictional B. strong
   C. gravitational D. unbalanced

11. A force times the distance through which the force moves is a definition of 11.___
   A. effort B. resistance
   C. friction D. work

12. A person uses a lever with a mechanical advantage of four to move a five hundred pound mass. 12.___
The effort force applied by the man is
   A. increased by 4 B. reduced by 4
   C. increased by $\frac{1}{4}$ D. decreased by $\frac{1}{4}$

13. The ideal mechanical advantage of a machine is ____ the actual mechanical advantage. 13.___
   A. less than B. equal to
   C. more than D. multiplied by

14. Simple machines are mechanical devices used to multiply force, change the direction of a force or gain speed. 14.___
An example of a machine used MAINLY to change the direction of a force is a
   A. screwdriver B. baseball bat
   C. fixed pulley D. wedge

15. That point on which a lever is supported and is free to rotate is called the 15.___
   A. effort B. resistance
   C. moment D. fulcrum

16. Anywhere in the universe, matter will ALWAYS have 16.___
   A. solid form B. weight C. life D. mass

17. ____ force causes falling bodies to accelerate. 17.___
   A. Frictional B. Gravitational
   C. Internal D. Balanced

18. A car travelling at 60 km in two hours has an AVERAGE speed of 18.___
   A. 30 km/2 hr. B. 30 km/hr
   C. 30 km/min D. 30 km/sec

19. A vector shows magnitude and 19.___
   A. mass B. direction C. weight D. velocity

20. The rate at which an object moves in a given direction is called 20.___
   A. speed B. motion C. magnitude D. velocity

21. Ball *A* is thrown horizontally. At the same instant, Ball *B* is dropped from the same height. 21.___
When the two balls strike the ground,
   A. ball *A* lands first
   B. ball *B* lands first
   C. they both hit the ground at the same time
   D. they both hit the ground in the same place

22. An arrow released from a bow has GREAT momentum because of its 22.___
   A. mass B. weight
   C. shape D. mass and velocity

23. The recoil of a gun is an example of 23.___
   A. Newton's First Law of Motion
   B. Newton's Second Law of Motion
   C. Newton's Third Law of Motion
   D. locomotion

Questions 24-28.

DIRECTIONS: Match the item at the right with the CORRECT choice on the left.

24. Subatomic particles with a negative charge 24.___
25. Have same number of protons, different number of neutrons 25.___
26. Electrons are transferred 26.___
27. Positive subatomic particle in nucleus 27.___
28. Nuclear particle with no charge 28.___

A. Isotopes
B. Electron
C. Protons
D. Neutrons
E. Ionic bond

29. Substances which CANNOT be broken down by ordinary physical or chemical means are 29.___
   A. elements B. mixtures
   C. gases D. compounds

30. In the figure at the right, the atomic number of Na is 30.___
   A. 11
   B. 22.98
   C. 281
   D. 182

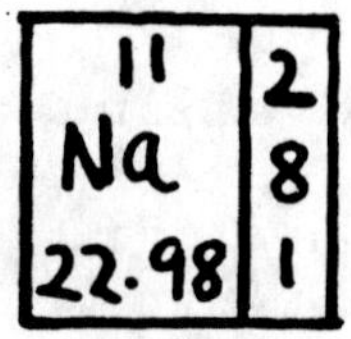

31. The symbol Na in the figure in the previous question represents 31.___
   A. nickel B. nitrogen C. sodium D. salt

32. In the figure in Question 30, the element Na contains ____ electrons. 32.___
   A. 11 B. 12 C. 281 D. 22.98

33. The first electron shell outside the nucleus contains ____ electron(s). 33.___
   A. 1 B. 2 C. 8 D. 12

34. An object which has a definite shape and definite volume is a 34.___
A. solid B. liquid C. gas D. plasmid

35. A certain block has a volume of 5 cm^3 and a mass of 4 grams. 35.___
The density is _____ gm/cm^3.
A. 0.5 B. 0.8 C. 8.0 D. 4.0

KEY (CORRECT ANSWERS)

1. A	11. D	21. C	31. C
2. C	12. C	22. D	32. A
3. D	13. A	23. C	33. B
4. A	14. C	24. B	34. A
5. B	15. D	25. A	35. B
6. B	16. D	26. E	
7. A	17. B	27. C	
8. C	18. B	28. D	
9. A	19. B	29. A	
10. D	20. D	30. A	

TEST 2

DIRECTIONS: Each question or incomplete statement is followed by several suggested answers or completions. Select the one that BEST answers the question or completes the statement. *PRINT THE LETTER OF THE CORRECT ANSWER IN THE SPACE AT THE RIGHT.*

1. Blue litmus paper turns ____ in acid. 1.___
   A. red B. orange C. yellow D. green

2. In a chemical change, the total mass 2.___
   A. increases B. remains the same
   C. decreases D. grows

3. If you want to increase the amount of sugar that dissolves in coffee, you should 3.___
   A. cool the coffee
   B. heat the coffee
   C. evaporate the coffee
   D. let the coffee stand overnight

4. Heat can travel by radiation, conduction, and convection. Only one of these methods of transport will allow heat to travel through a vacuum. 4.___
   This method is
   A. conduction B. convection
   C. radiation D. none of this occurs

5. 50 ml of water at 60°C are mixed with 50 ml of water at 30°C. 5.___
   The resulting liquid now has a temperature of ____ °C.
   A. 60 B. 30 C. 45 D. 90

6. One calorie is 6.___
   A. the amount of fat on a person's body
   B. a unit of mass when dealing with food
   C. a unit of heat
   D. an enormous amount of energy

7. The wave length of the wave in the figure at the right is indicated by arrow 7.___
   A. A
   B. B
   C. C
   D. D

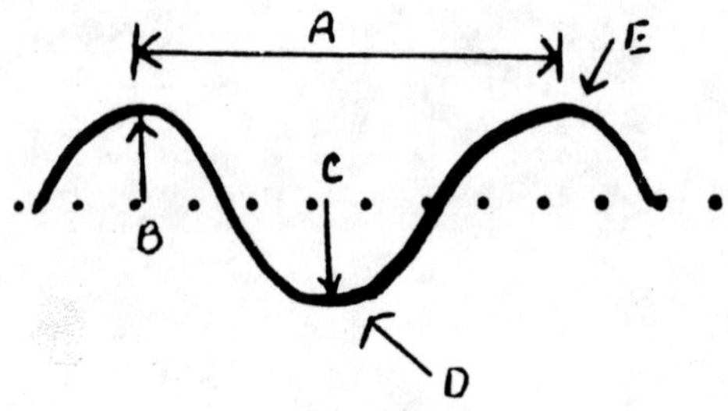

8. The crest of a wave is indicated in the figure in the previous question by arrow 8.___
   A. A B. B C. C D. D

9. The trough of a wave is indicated in the figure in Question 7 by arrow 9.___
   A. A B. B C. C D. D

10. The frequency of a wave means 10.___
   A. the size of the wave
   B. how often the wave occurs in a time period
   C. the velocity of the wave
   D. waves do not have frequency

11. An infrared wave is a(n) 11.___
   A. light wave B. heat wave
   C. x-ray D. cosmic wave

Questions 12-16.

DIRECTIONS: Match the item at the right with the CORRECT choice on the left.

12. Bending of light rays A. Reflection of light 12.___

13. Bouncing back light rays B. Refraction of light 13.___

14. A prism 14.___

15. A mirror 15.___

16. A microscope 16.___

17. An example of a convex lens is 17.___

A. 
B. 
C. 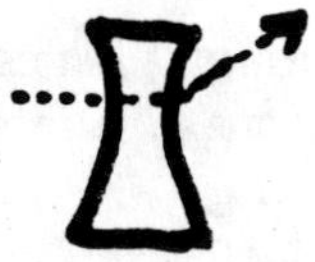
D. 

18. An example of a mirror is 18.___

A. 
B. 
C. 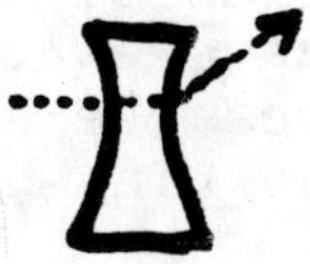
D. 

19. An example of a prism is 19.___

A. 
B. 
C. 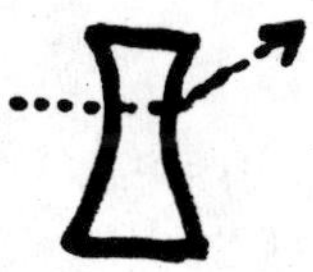
D. 

20. An example of a concave lens is 20.___

A. 
B. 
C. 
D. 

Questions 21-24.

DIRECTIONS: Match the item at the right with the CORRECT choice on the left.

21.	From lens to principal focus	A. Real image	21.___
		B. Retina	
22.	Can be projected	C. Farsightedness	22.___
23.	Light-sensitive lining	D. Focal length	23.___
24.	Weakness in near vision		24.___

25. Objects become electrically charged when they gain or lose 25.___
    A. protons B. neutrons C. electrons D. neutrinos

26. A pathway through which electric current travels is a 26.___
    A. galvanometer B. ammeter
    C. circuit D. volt meter

27. Electricity is a form of 27.___
    A. light B. force C. energy D. work

28. When the leaves of an electroscope stand apart, the leaves have 28.___
    A. the same charge on each leaf
    B. opposite charges
    C. no charge
    D. resistance

29. A swimming pool and a glass of water are both 9°C. It can be said that 29.___
    A. the swimming pool has more heat
    B. the glass of water has more heat
    C. both have equal amounts of heat
    D. it takes the same amount of time to heat the pool and the glass of water

30. Wire wrapped around a nail and connected to a battery will cause the nail to 30.___
    A. generate small discharges of small voltage electricity
    B. become magnetized
    C. glow
    D. rust

31. Nobel gases have ____ in their outermost shells. 31.___
    A. 6 electrons B. 8 electrons
    C. 8 protons D. 8 neutrons

32. A balance scale is used to measure 32.___
    A. volume B. mass C. gravity D. force

33. A scientist GENERALLY repeats experiments 33.___
A. because he makes mistakes
B. because it is fun
C. to reduce errors in chance
D. true scientists do an experiment only once

34. A compressed spring represents 34.___
A. kinetic energy B. pep
C. potential energy D. elasticity

35. A charged particle is called a(n) 35.___
A. atom B. molecule C. ion D. isotope

KEY (CORRECT ANSWERS)

1. A	11. B	21. D	31. B
2. B	12. B	22. A	32. B
3. B	13. A	23. B	33. C
4. C	14. B	24. C	34. C
5. C	15. A	25. C	35. C
6. C	16. B	26. C	
7. A	17. B	27. C	
8. B	18. D	28. A	
9. C	19. A	29. A	
10. B	20. C	30. B	

EXAMINATION SECTION

TEST 1

DIRECTIONS: Each question or incomplete statement is followed by several suggested answers or completions. Select the one that BEST answers the question or completes the statement. *PRINT THE LETTER OF THE CORRECT ANSWER IN THE SPACE AT THE RIGHT.*

1. The metric unit of length is the 1.____
   A. pound B. foot C. gram D. kilogram E. meter

2. What basic unit of time is common to ALL measurement systems? 2.____
   A. Hour B. Week C. Day
   D. Second E. Fortnight

3. If length in the *cgs* division of the metric system is measured in centimeters (abbreviated cm), then volume in that same system would be measured in 3.____
   A. cm^3 B. cm^2 C. cm D. quarts E. pints

4. ____ centimeters are in one meter. 4.____
   A. 0.1 B. 1 C. 10 D. 100 E. 1000

5. ____ grams are in one kilogram. 5.____
   A. 1 B. 10 C. 100 D. 1000 E. 1,000,000

6. The measure of the amount of matter in an object is called the object's 6.____
   A. volume B. density C. mass
   D. weight E. solubility

7. ____ comes CLOSEST to being your mass in kilograms. 7.____
   A. 500 B. 50 C. 5 D. 0.5 E. 0.05

8. A cube of steel is one cm on each edge. Its mass is 5.6 grams (abbreviated g). 8.____
   What is the density of this steel cube?
   A. 0.56 g/cm B. 0.56 g/cm^2 C. 5.6 g/cm^3
   D. 5.6 g E. 56 g/cm^3

9. Density is a(n) ____ property of matter. 9.____
   A. physical B. chemical C. immeasurable
   D. positive E. electric

10. The fact that a piece of magnesium burns in air with a brilliant white flame is an example of a(n) ____ property. 10.____
    A. physical B. chemical C. electric
    D. potential E. impossible

Questions 11-13.

DIRECTIONS: In Questions 11 through 13, you are asked about the important idea of classes of matter. Answer in this series of questions:
A. Element
B. Compound
C. Mixture
D. None of these three

11. Your breakfast bowl of corn flakes, sugar, and milk 11.___

12. The salt on the table (sodium chloride) 12.___

13. The oxygen in the air you breathe (the oxygen, not the air) 13.___

14. The SMALLEST part of an element is called a(n) 14.___
A. mixture B. fundamental C. proton
D. atom E. none of the above

15. The SMALLEST part of a compound is called a(n) 15.___
A. atom B. molecule C. proton
D. mixture E. none of the above

16. When elements combine together chemically, they form 16.___
A. mixtures B. elements C. reactions
D. compounds E. explosions

Questions 17-20.

DIRECTIONS: In Questions 17 through 20, you are to identify the common elements whose abbreviations are given in the list to the right. Note that NOT ALL of the answers will be used.

17. Oxygen A. Fe 17.___

18. Carbon B. O 18.___

19. Iron C. S 19.___

20. Sodium D. C 20.___

E. Na

21. In the compound potassium nitrate, each group or molecule contains how many TOTAL atoms? (The formula for potassium nitrate is KNO_3.) 21.___
A. 3 B. 5 C. 6
D. 9 E. An incalculable number

Questions 22-25.

DIRECTIONS: Questions 22 through 25 refer to the equation which is written below.

$$KI + Pb(NO_3)_2 \rightarrow KNO_3 + PbI_2$$

22. Which of the materials in this equation would be called products? 22.____
   A. KI
   B. KNO_3 *only*
   C. KNO_3 and KI
   D. KNO_3 and PbI_2
   E. $Pb(NO_3)_2$ and PbI_2

23. How many nitrogen (N) atoms are present in each molecule of $Pb(NO_3)_2$? 23.____
   A. None B. One C. Two D. Five E. Six

24. ____ is the CORRECT name for the compound PbI_2. (Pb is lead; I is iodine.) 24.____
   A. Lead iodide
   B. Lead iodine
   C. Iodine lead
   D. Leaded iodine
   E. Iodic lead

25. As the reaction is written above, is it chemically balanced? 25.____
   A. Yes, it is balanced.
   B. No, it is not balanced.
   C. It is impossible to tell if it is balanced.
   D. It is not even a reaction; it could not be balanced.
   E. Reactions are never balanced.

26. Two liquids in two separate beakers stand on the pan of a balance which reads the mass of everything on it. One of the beakers is picked up, its contents poured into the other beaker, and then set back down on the balance again. As the two liquids mix, a bright orange precipitate forms. What will be the NEW mass of the liquids and beakers? 26.____
   A. Less than before
   B. More than before
   C. The same as before
   D. Impossible to predict

27. Elements which lie in the SAME vertical column (same group) on the periodic table have 27.____
   A. similar numbers of particles
   B. similar chemical properties
   C. the same mass
   D. equal charges
   E. the same father

Questions 28-30.

DIRECTIONS: Questions 28 through 30 ask you some questions about the subatomic particles. Answer:
A. Electrons
B. Protons
C. Neutrons
D. Electrons and protons
E. Protons and neutrons

28. Which particle or particles possess(es) electric charge? 28.___

29. Which particle or particles is(are) found in the inside of the atom's nucleus? 29.___

30. Which particle is associated with the identity of the element (that is, the number of this particle establishes which element it is)? 30.___

KEY (CORRECT ANSWERS)

1. E	11. C	21. B
2. D	12. B	22. D
3. A	13. C	23. C
4. D	14. D	24. A
5. D	15. B	25. B
6. B	16. D	26. C
7. B	17. B	27. B
8. C	18. D	28. D
9. A	19. A	29. E
10. B	20. E	30. B

TEST 2

DIRECTIONS: Each question or incomplete statement is followed by several suggested answers or completions. Select the one that BEST answers the question or completes the statement. *PRINT THE LETTER OF THE CORRECT ANSWER IN THE SPACE AT THE RIGHT.*

1. The family of elements making up the far right column on the periodic table includes helium, neon, argon, krypton, xenon, and radon. These elements are known as the inert gases and have 1.____
   A. very dense nuclei
   B. no electrons in their orbits
   C. low atomic numbers
   D. very stable electron arrangements
   E. none of the above

2. The element immediately to the left of neon (one of the inert gases) on the periodic table is fluorine. We know from this, then, that fluorine has a 2.____
   A. strong tendency to gain an electron
   B. strong tendency to lose an electron
   C. weak tendency to gain an electron
   D. weak tendency to lose an electron
   E. tendency to neither gain nor lose electrons

3. When magnesium forms compounds, it loses two electrons. What is the ion of magnesium thus formed? 3.____
   A. Mg B. Mg^{+1} C. Mg^{-1} D. Mg^{+2} E. Mg^{-2}

4. If magnesium described in the previous question reacts with chlorine, which has a strong tendency to gain one electron, what would be the PROPER formula for the compound magnesium chloride? 4.____
   A. MgCl B. Mg_2Cl_2 C. $MgCl_2$
   D. Mg_2Cl E. None of the above

5. The hydroxides of metals, like NaOH, KOH, LiOH for example, are all classed as 5.____
   A. acids B. bases C. salts
   D. lipids E. metalloids

6. When an acid reacts with a base, resulting in water and a salt as products, the reaction is called 6.____
   A. substitution B. decomposition C. synthesis
   D. neutralization E. fertilization

7. What man is credited with the arrangement of the elements into the structure we call the periodic table? 7.____
   A. Mendeleev B. Rutherford C. Dalton
   D. Lavoisier E. Compton

8. The following three nuclides, ${}^{12}_{6}C$, ${}^{13}_{6}C$, ${}^{14}_{6}C$, are called 8.___

A. ions B. cations C. isotopes
D. radioactive E. isomers

9. The nuclide ${}^{14}_{6}C$ is useful in medicine and biology because it decays from its nucleus, ejecting a beta particle during that decay. 9.___
We say that it is
A. dangerous B. an ion C. an anion
D. radioactive E. stable

10. When a chemist talks about *organic compounds*, he is talking about a wide range of materials. 10.___
One thing they have in common, however, is that they ALL contain
A. oxygen B. phosphorous C. sulfur
D. carbon E. life

11. Nitrogen, phosphorous, and sulfur are important 11.___
A. organic elements B. metals
C. ions D. nonmetals
E. compounds

12. Synthetic rubber and plastic, as well as cellulose in plants, are all examples of long molecules with many repeating units. 12.___
This classifies them as
A. sugars B. useless C. polymers
D. polyglots E. pollywannacrackers

13. When we warm up a substance, its particles (molecules, atoms, ...) become 13.___
A. more charged B. stickier C. faster moving
D. impatient E. less reactive

14. In order to change the phase (state) of a material from a solid to a liquid, we would have to 14.___
A. add heat B. remove heat C. add cold
D. remove cold E. ionize it

15. The amount of heat needed to warm up a 10 gram sample of water from 20°C to 25°C would be ____ calorie(s). 15.___
A. 1 B. 10 C. 5 D. 50 E. 500

16. *Generally*, heating a substance causes that substance to expand. Thus, warm water floats on top of layers of colder water. 16.___
This is because
A. the water is warmer
B. the warmer water has a lower density
C. heat rises
D. slower molecules sink to the bottom
E. warmer water has a greater density

17. What is normal body temperature for a person, expressed in °C? 17.____
 A. 98.6 B. 87 C. 68 D. 20 E. 37

18. A car travels at an average speed of 20 meters/second. The car will travel ____ meters in one minute. 18.____
 A. 20 B. 200 C. 1200 D. 6000 E. 12,000

19. A car accelerates from rest at a uniform rate so that 11 seconds after it started, it is travelling at 55 miles per hour. 19.____
What is the car's rate of acceleration during this time?
 A. 55 miles/hour
 B. 55 miles/second
 C. 55 miles/hour/second
 D. 5 miles/second/hour
 E. 5 miles/hour/second

20. The car in the previous question continues at this rate of acceleration for another 2 seconds after the 11 seconds indicated in the problem. 20.____
How fast will it be going at the end of this additional time?
 A. 5 miles/hour/second
 B. 65 miles/hour
 C. 55 miles/second
 D. 65 miles/hour/second
 E. Impossible to predict

21. According to Newton's Second Law of Motion, ____ causes an object to accelerate. 21.____
 A. mass
 B. momentum
 C. torque
 D. friction
 E. force

22. Remember that gravity causes objects to accelerate at a constant rate when they fall near the surface of the earth. This acceleration is nearly (but NOT quite) equal to 10 m/second^2. 22.____
How fast would a dropped stone be falling after three seconds of fall near the Earth's surface?
 A. 10 m/second
 B. 10 m/second^2
 C. 30 m/second
 D. 30 m/second^2
 E. 300 m/second

23. The resistance that an object has to being accelerated, described in Newton's First Law of Motion, is 23.____
 A. momentum
 B. inertia
 C. torque
 D. friction
 E. none of the above

24. When you fire a gun, the gun kicks back. 24.____
According to Newton's Third Law of Motion,
 A. the force of the gun on the bullet equals the force of the bullet on the gun
 B. the force of the gun on the bullet is the only force in the system
 C. the force of the bullet on the gun is the only force in the system
 D. the forces of the gun and the bullet cancel each other out
 E. no force is required to shoot the bullet from the gun

25. The ability in a system to do work or to cause change is 25.___
A. efficiency B. heat C. conservation
D. energy E. entropy

Questions 26-28.

DIRECTIONS: In Questions 26 through 28, indicate the KIND OF ENERGY present in the system described by the question. Answer:
A. Kinetic energy
B. Thermal energy (heat)
C. Potential gravitational energy
D. Mechanical energy (work)
E. Chemical energy

26. The energy possessed by Superman because he is moving faster than a speeding bullet. 26.___

27. The energy possessed by a truck sitting stationary at the top of a long, high hill. 27.___

28. The kind of energy available to you in your breakfast bowl of Rice Krispies, milk, and sugar you ate this morning. 28.___

29. You have to spend 8000 units of energy to run your air conditioner for one hour, but you only pump out 4000 units of energy as heat from your room while the air conditioner is running. 29.___
What is the air conditioner's efficiency?
A. 100% B. 400% C. 40% D. 25% E. 5%

30. The pitch of a sound is related MOST closely to the sound's 30.___
A. loudness B. waves C. amplitude
D. frequency E. source

31. The color of visible light is related MOST closely to the light's 31.___
A. wave length B. reflections C. strength
D. absorbtion E. intensity

32. A red jacket looks red to your eye because it 32.___
A. reflects all colors but red
B. absorbs all colors but red
C. transmits all colors but blue
D. reflects only blue
E. absorbs only red

33. A material which is a good insulator MUST also be a 33.___
A. poor conductor
B. good conductor
C. metal
D. good electron source
E. good acceptor of electrons

34. The rate at which electric charge flows in a circuit is BEST termed 34.___
A. voltage
B. charge
C. pressure
D. current
E. power

35. When an electric current flows through a wire which has resistance, the result is that the wire consumes electric energy. 35.___
The MOST noticeable result of this power consumption is that
A. the wire is charged
B. the wire gets warm
C. electrons escape from the wire
D. the battery charges up
E. the wire changes colors

KEY (CORRECT ANSWERS)

1. D	11. D	21. B	31. A
2. A	12. C	22. D	32. B
3. D	13. C	23. B	33. A
4. C	14. A	24. A	34. A
5. B	15. C	25. D	35. B
6. D	16. B	26. D	
7. C	17. E	27. C	
8. C	18. C	28. E	
9. D	19. E	29. E	
10. D	20. B	30. D	

EXAMINATION SECTION

TEST 1

DIRECTIONS: Each question or incomplete statement is followed by several suggested answers or completions. Select the one that BEST answers the question or completes the statement. *PRINT THE LETTER OF THE CORRECT ANSWER IN THE SPACE AT THE RIGHT.*

1. One advantage of using the metric system is that 1.___
   A. the units are all related by powers of ten
   B. you cannot move decimal points to change from one unit to another
   C. the units are NOT related in any way
   D. the units may be easily converted to English units

2. A carpenter measures a board four times and finds the values 40 cm, 41 cm, 41 cm, 42 cm. 2.___
   The average of his measurements is ____ cm.
   A. 40 B. 41 C. 42 D. 4

3. An object with a mass of 1 gram ____ weight. 3.___
   A. must have B. may have
   C. has no D. cannot have

4. Your speed if you travelled 990 kilometers in three hours was ____ kilometers per hour. 4.___
   A. 330 B. 3 C. 660 D. 165

5. Which of these graphs BEST represents the relationship between the object placed on a spring scale and the amount the spring stretches? 5.___

A.

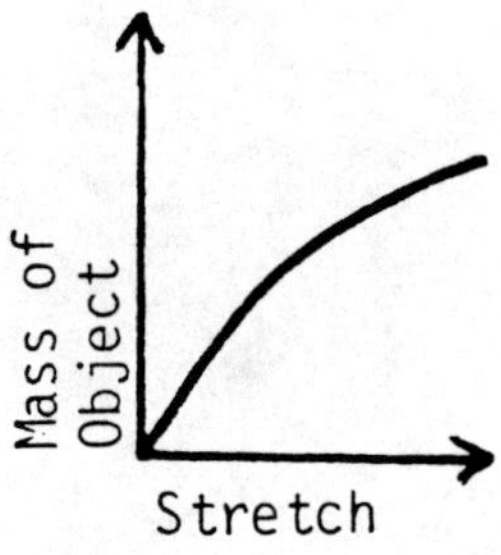

B.

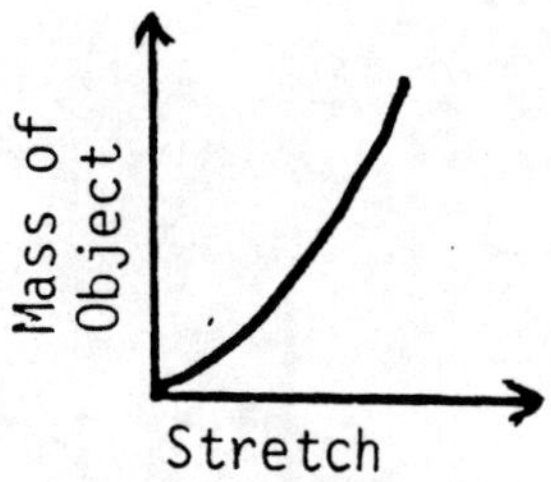

C.

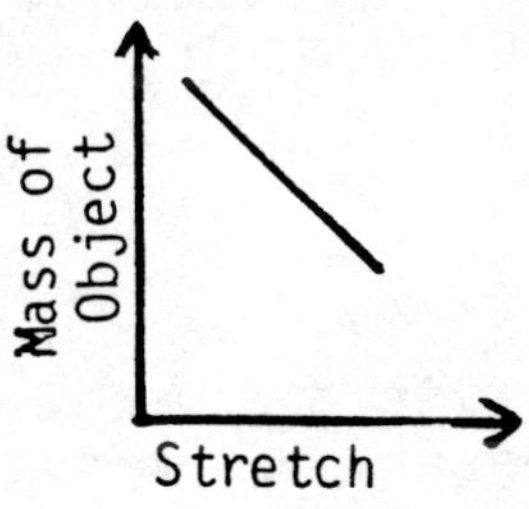

D.

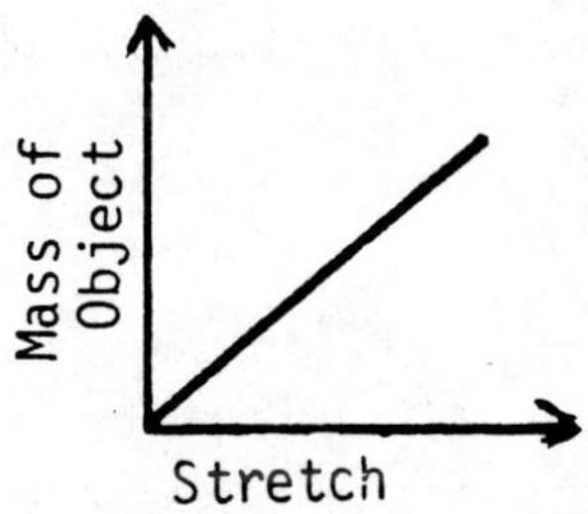

6. An example of acceleration is 6.___
   A. -10 Km/min B. 10 Km/hr
   C. 10 Km/hr/hr D. 10 Km/Km/sec

7. The metric unit of length is the 7.___
   A. liter B. meter C. gram D. foot

8. The gravitational pull of the earth on a body is the 8.___
   A. weight B. mass C. velocity D. acceleration

9. A student in ninth grade might *likely* have a height of 9.___
   A. 1 meter B. 3 meters C. 500 cm D. 150 cm

10. A cannon ball is taken to the moon. 10.___
    Compared to its weight on earth, its weight on the moon will be
    A. increased B. unchanged
    C. decreased D. none of the above

11. The same cannon ball as in the previous question is taken to the moon. 11.___
    Its mass will be
    A. increased B. decreased
    C. remains the same D. none of the above

12. Which graph shows constant velocity? 12.___

A.

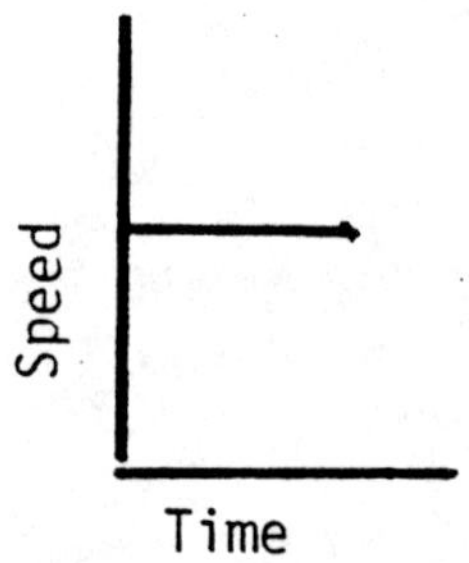

B.

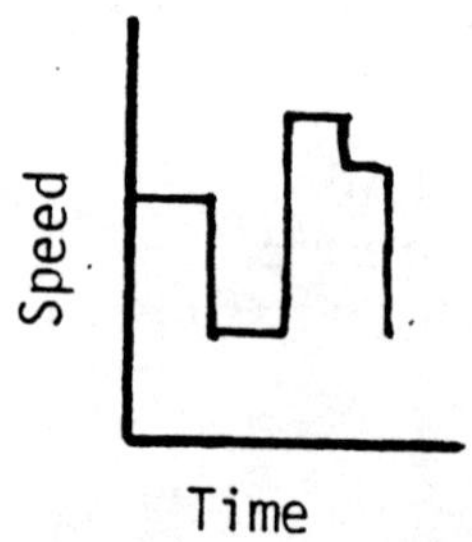

C.

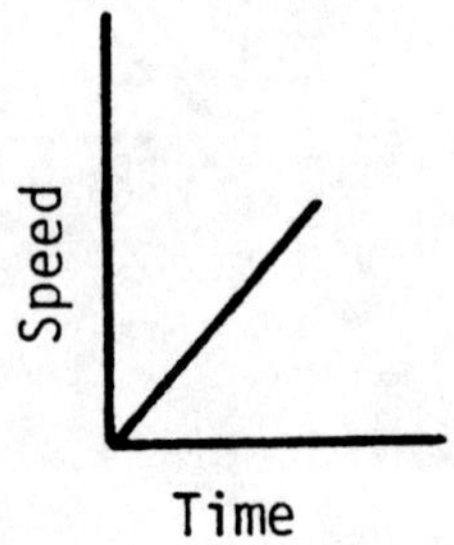

D.

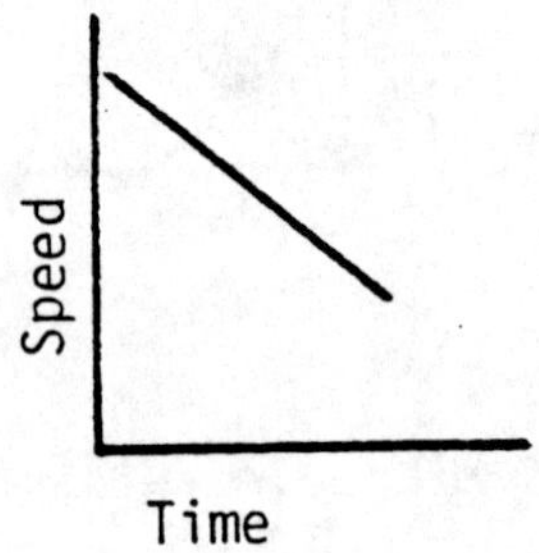

13. Which of these statements is CORRECT? 13.___
    A. It takes more force to maintain a steady velocity.
    B. It takes more force to cause an acceleration.
    C. It takes no force to cause an acceleration.
    D. Force is NOT related to motion.

14. Travelling at 50 miles per hour, a ____ will have more momentum. 14.___

A. Truck B. Train
C. Little red wagon D. Car

15. A boy is sitting in a chair. He weighs 100 pounds. According to Newton's Third Law of Motion, the 15.___

A. chair will accelerate downward
B. boy will break the chair
C. chair pushes back with 100 pounds of force
D. boy is doing 100 lbs. of work

16. More work is done when lifting a box with 16.___

A. a single fixed pulley
B. a single movable pulley
C. two movable pulleys
D. neglecting friction, the amount of work done is the same for each of the above examples

17. The volume of gasoline tank can be expressed in 17.___

A. millimeters B. milliliters
C. cubic centimeters D. both B and C above

18. Two heavy objects of equal mass, one a bit larger than the other, are dropped at the same time from a high tower. Ignoring the effect of air friction, 18.___

A. the larger object will hit first
B. the smaller object will hit first
C. the heavier object will hit first
D. they both will hit the ground at the same time

19. Suppose you wanted to find out if some chemical affects the behavior of fish. You put the fish into a container of water and add some of the chemical. Then you look for changes in the behavior of the fish. 19.___
If the behavior of the fish does actually change, which conclusion could you reach?

A. The chemical affected the behavior of the fish.
B. The behavior of the fish caused a change in the chemical.
C. No accurate conclusion can be reached since the experiment lacks control.
D. No accurate conclusion can be reached since the experiment has NO variables.

20. The area of the triangle at the right is ____ cm^2. 20.___

A. 12
B. 6
C. 7
D. 1

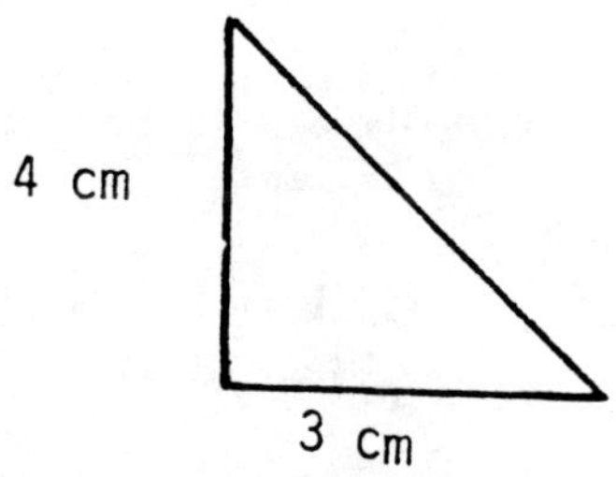

21. A machine is 100% efficient if 21.___

A. the ideal mechanical advantage equals the actual mechanical advantage
B. the machine can lift 100 pounds
C. work input is less than work output
D. work output is less than work input

22. An airplane has a cruising speed of 300 miles/hour and is bucking a 60 mile/hour headwind. In four hours, it can cover _____ miles. 22.____
 A. 960 B. 1400 C. 1800 D. 300

23. Indirect measurements use 23.____
 A. a microscope
 B. a scale drawing
 C. a telescope
 D. it is impossible to measure something you cannot touch

24. The central core of an atom is the 24.____
 A. neutron B. nucleus C. orbit D. electron

25. An atom has 17 protons, 17 electrons, and 18 neutrons. Its atomic weight is 25.____
 A. 17 B. 34 C. 35 D. 52

26. The halogens will enter into chemical reactions by gaining electrons because they have _____ outer shell. 26.____
 A. 7 electrons in the
 B. 1 electron in the
 C. a complete
 D. various combinations of electrons in the

27. Ionic compounds are formed by the transfer of 27.____
 A. neutrons B. protons
 C. electrons D. all of the above

28. _____ and _____ are combinations of ions which unite chemically to become common table salt. 28.____
 A. Sodium; nitrogen B. Nitrogen; oxygen
 C. Sodium; chlorine D. Nitrogen; chlorine

29. Which of these atomic electron arrangements requires one electron to become stable? 29.____
 A. 2,8 B. 2, 7 C. 2, 6 D. 2, 8, 1

30. Which of these elements is a noble gas? 30.____
 A. Carbon B. Nitrogen C. Helium D. Oxygen

31. An element of the halogen family is 31.____
 A. chlorine B. oxygen
 C. nitrogen D. alkali metal

32. Sulfur is in the same family as 32.____
 A. chlorine B. oxygen
 C. nitrogen D. alkali metal

33. All hydrocarbons contain 33.____
 A. chlorine B. carbon C. oxygen D. helium

34. _____ can be converted to heat energy. 34.____
 A. Electricity B. Electromagnetic radiation
 C. Chemical energy D. All of the above

35. A turbine is a kind of 35.____
 A. water dam B. water wheel
 C. shaft D. A.C. converter

36. Modern hydropower uses ____ and ____. 36.____
 A. turbines; generators B. turbines; batteries
 C. generators; shafts D. none of the above

Questions 37-41.

DIRECTIONS: In answering Questions 37 to 41, match the description in Column II with the term it describes in Column I.

Column I	Column II	
37. Inert	A. A charged particle	37.____
38. Noble gas	B. An electron arrangement of completely filled shells	38.____
39. Chemical activity		39.____
40. Stable arrangement	C. A gas having a stable electron arrangement	40.____
41. Ion	D. The way an atom reacts with other atoms	41.____
	E. Not chemically active	

42. Which scientist formulated the modern periodic law? 42.____
 A. Mendeleev B. Dalton C. Moseley D. Einstein

43. Which of these is arranged vertically in the periodic table? 43.____
 A. Electrons B. Protons
 C. Neutrons D. Groups (families)

44. ____ proposed an atomic theory. 44.____
 A. Mendeleev B. Dalton C. Moseley D. Einstein

45. ____ arranged the information vertically in the periodic table. 45.____
 A. Mendeleev B. Dalton C. Moseley D. Einstein

46. Which of these can be considered a different form of an element? 46.____
 A. Electron B. Isotopes C. Neutron D. Proton

47. Which BEST describes the way heat moves through a solid? 47.____
 A. Convection B. Conduction C. Osmosis D. Radiation

48. Which BEST describes the way heat moves through liquids and gases? 48.____
 A. Convection B. Conduction C. Osmosis D. Radiation

49. Water boils at 100 degrees 49.___
   A. Fahrenheit
   B. Kelvin
   C. Celsius
   D. water does not boil at 100°

50. ____ is the source of 90% of the heat of the earth. 50.___
   A. The sun's radiant energy
   B. Volcanic activity
   C. Radioactivity
   D. The moon

KEY (CORRECT ANSWERS)

1. A	11. C	21. D	31. A	41. A
2. B	12. A	22. A	32. B	42. A
3. B	13. B	23. A	33. B	43. D
4. A	14. B	24. B	34. D	44. B
5. D	15. C	25. C	35. B	45. A
6. C	16. D	26. A	36. A	46. B
7. B	17. B	27. C	37. E	47. B
8. A	18. D	28. C	38. C	48. A
9. C	19. A	29. B	39. D	49. C
10. C	20. B	30. C	40. B	50. A

TEST 2

DIRECTIONS: Each question or incomplete statement is followed by several suggested answers or completions. Select the one that BEST answers the question or completes the statement. *PRINT THE LETTER OF THE CORRECT ANSWER IN THE SPACE AT THE RIGHT.*

1. Which one of these measures is dependent on matter's average kinetic energy? 1.___
   A. Mass B. Volume C. Temperature D. Time

Questions 2-6.

DIRECTIONS: In answering Questions 2 to 6, match from the left column the type of energy which BEST describes the object in the right column.

2. A moving truck	A. Kinetic energy	2.___
3. A lump of coal	B. Potential energy	3.___
4. Water stored behind a dam		4.___
5. A stick of dynamite		5.___
6. A falling weight		6.___

Questions 7-9.

DIRECTIONS: In answering Questions 7 to 9, use the following figure.

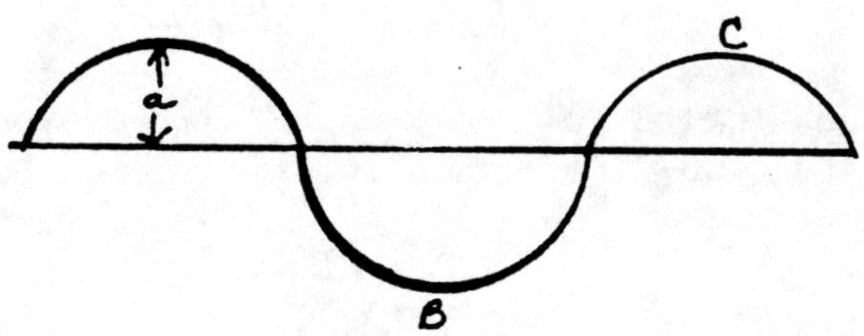

7. In the wave diagram above, part *a* is the ____ of the wave. 7.___
   A. crest B. trough C. wavelength D. amplitude

8. In the wave diagram above, part *b* is the ____ of the wave. 8.___
   A. crest B. trough C. wavelength D. amplitude

9. In the wave diagram above, part *c* is the ____ of the wave. 9.___
   A. crest B. trough C. wavelength D. amplitude

10. Rhythmical disturbances that travel through air, water, and metal are called 10.___
A. springs B. hills C. waves D. valleys

11. A slinky can be used to demonstrate ____ 11.___
A. transverse waves B. longitudinal waves
C. both A and B above D. none of the above

12. Sound waves move through 12.___
A. solids, liquids, gases B. space *only*
C. air *only* D. space and solids

13. To calculate the speed of a wave, you must know 13.___
A. the amplitude and frequency
B. the frequency and wavelength
C. the wavelength and amplitude
D. none of the above

14. The faster a string vibrates, the higher is the 14.___
A. pitch B. frequency
C. neither A nor B D. both A and B

15. Sounds produced by irregular vibrations are called 15.___
A. music B. highs and lows
C. noise D. singing

16. The electromagnetic spectrum has 16.___
A. a region of visible light
B. radio waves and cosmic waves
C. x-rays and heat waves
D. all of the above

17. The velocity of a wave changes as it 17.___
A. passes from one density to another
B. passes through a single density
C. moves farther from its source
D. the velocity never changes

18. The law which states *the angle of incidence equals the angle of reflection* is the law of 18.___
A. interference B. refraction
C. reflection D. light

19. The spot in front of a curved mirror where the light is reflected is the 19.___
A. ray B. center of curvature
C. principal axis D. principal focus

20. An image formed by a concave mirror can be 20.___
A. larger than the object
B. smaller than the object
C. same size as the object
D. all of the above

21. An object that allows all the light that strikes it to be transmitted is 21.___
    A. opaque B. transparent
    C. luminous D. illuminated

22. A fuzzy shadow is created by a 22.___
    A. point source of light B. large light source
    C. any light source D. the sun

23. A ray of light that enters a piece of glass at 0° will 23.___
    A. bend B. reflect back
    C. not bend D. be absorbed and disappear

24. We see objects as colored because of 24.___
    A. refraction B. diffraction C. reflection D. diffusion

25. Potassium has a valence of +1. Oxygen has a valence of -2. 25.___
    The chemical formula for potassium oxide is
    A. KO B. K_2O C. KO_2 D. none of these

26. When a band is formed between two atoms, the electrons in the ____ shell are ALWAYS involved. 26.___
    A. K B. L C. inner D. outer

27. Metals are good conductors of electricity because they are 27.___
    A. usually gray
    B. usually solid
    C. in possession of free electrons
    D. good insulators

28. As temperature increases, the motion of particles in a liquid 28.___
    A. ALWAYS increases B. ALWAYS decreases
    C. stays the same D. sometimes decreases

29. A block of aluminum has a volume of 20 cm^3 and a mass of 54 g. 29.___
    The density of aluminum is ____ g/cm^3.
    A. 2.7 B. .37 C. 10.80 D. 1080
    D.

30. The block of aluminum in the previous question is cut in half. 30.___
    The density of each half is
    A. the same as the original density
    B. 1.35 g/cm^3
    C. 5.4 g/cm^3
    D. not enough information

Question 31.

DIRECTIONS: In answering Question 31, refer to the chart below.

Substance	Density g/cm^3
Oak	.65
Alcohol	.79
Ice	.92
Water	1.00
Iron	7.80
Copper	8.90
Mercury	13.60

31. Referring to the above chart, which of these statements is FALSE? 31.___
    A. Mercury has the greatest density.
    B. Oak will float on water.
    C. Iron will float on alcohol.
    D. Copper will float on mercury.

32. During a physical change, 32.___
    A. a new substance is formed
    B. there is a large energy change
    C. a change of size cannot occur
    D. the original substances remain

33. When a substance is boiling, 33.___
    A. the temperature goes up
    B. the temperature remains the same
    C. it is turning into a liquid
    D. both A and C above

34. To make fresh water from salt water in one step, you could use 34.___
    A. distillation B. filtration
    C. electrolysis D. none of the above

35. Matter 35.___
    A. takes up space B. has mass
    C. has volume D. all of the above

36. One bulb burns out on a set of Christmas tree lights, and the rest of the lights go out. 36.___
    The light set is a ____ circuit.
    A. closed B. direct C. series D. parallel

37. An alternating current is produced by 37.___
    A. a generator B. a galvanometer
    C. a battery D. all of the above

38. You buy a radio which MUST be plugged into a wall socket. The radio is rated at 110 watts. You play the radio for 20 hours. 38.___
    The electric company will charge you ____ KWH.
    A. 5.5 B. 5500 C. 2200 D. 2.2

39. Electric current is the flow of 39.___
   A. protons B. neutrons C. electrons D. ions

40. Wire can increase its ability to give off heat by 40.___
   A. increasing the wire's insulation
   B. increasing the wire's resistance
   C. decreasing the voltage in the wire
   D. decreasing the wattage in the wire

Questions 41-45.

DIRECTIONS: In answering Questions 41 to 45, from Column II, select the substance which is described in Column I.

COLUMN I	COLUMN II	
41. Used to identify substances by changing colors	A. Acids	41.___
	B. Base	
42. Turns blue litmus paper red	C. Salts	42.___
	D. Solvent	
43. Formed in neutralization process	E. Indicators	43.___
44. Turns red litmus paper blue		44.___
45. Substance in which a solute will dissolve		45.___

46. Solar energy is a saving way to heat homes because 46.___
   A. the sun is free
   B. the sun rises every day
   C. the heat from the sun can be collected and saved to be used at night or on a cloudy day
   D. all of the above

47. An example of a solid solution is 47.___
   A. alloy B. metalloid
   C. sugar crystals D. iron

48. If 10g of salt are placed in a beaker with 15g of water, some salt will dissolve and some does NOT dissolve. What should be done to make all the salt dissolve? 48.___
   A. Hold the beaker in the light
   B. Place an ice cube in the beaker
   C. Add more water
   D. None of the above

49. Scientists predict that cars will soon be running on gasohol. 49.___
Gasohol is a mixture of gasoline and
A. water B. radioactivity C. alcohol D. acid

50. Nuclear generating stations are considered hazardous by some people because of 50.___
A. atomic explosions
B. radiation leaks into the atmosphere
C. their expensive operating costs
D. none of the above

KEY (CORRECT ANSWERS)

1. C	11. C	21. B	31. C	41. E
2. A	12. A	22. B	32. D	42. A
3. B	13. B	23. C	33. B	43. C
4. B	14. D	24. B	34. A	44. B
5. B	15. C	25. B	35. D	45. D
6. A	16. D	26. D	36. C	46. A
7. A	17. A	27. C	37. A	47. A
8. B	18. B	28. A	38. D	48. C
9. A	19. D	29. A	39. C	49. C
10. C	20. D	30. A	40. B	50. D

EXAMINATION SECTION

TEST 1

DIRECTIONS: Each question or incomplete statement is followed by several suggested answers or completions. Select the one that BEST answers the question or completes the statement. *PRINT THE LETTER OF THE CORRECT ANSWER IN THE SPACE AT THE RIGHT.*

1. The study of matter and energy is 1.___
   A. biology B. physical science
   C. geology D. mathematics

2. A scientific *guess* is a(n) 2.___
   A. law B. theory C. observation D. hypothesis

3. All of the following are forms of matter EXCEPT 3.___
   A. water B. air C. iron D. electricity

4. ____ cubic centimeters are equal to one milliliter. 4.___
   A. 1 B. 10 C. 100 D. 1000

5. Work is the product of force and 5.___
   A. time B. power C. distance D. speed

6. In a wheel and axle, how does the radius of the wheel compare with the radius of the axle? 6.___
   A. They are the same. B. The wheel is larger.
   C. The axle is larger.

7. ____ is a compound machine. 7.___
   A. Wheel and axle B. Screw
   C. Inclined plane D. Bicycle

8. Crowbars, scissors, and seesaws are examples of 8.___
   A. screws B. wedges
   C. levers D. inclined planes

9. Reducing friction can increase a machine's 9.___
   A. efficiency B. work input
   C. effort distance D. resistance arm

10. The newton is a unit of 10.___
    A. mass B. force C. inertia D. density

11. The standard metric unit of weight is the 11.___
    A. watt B. kilogram C. joule D. newton

12. The distance an object moves per unit of time is 12.___
    A. velocity B. acceleration
    C. speed D. displacement

13. The amount of kinetic energy an object has depends on the object's velocity and 13.____
A. magnetism B. volume C. weight D. mass

14. A stone resting on the edge of a cliff has NO 14.____
A. mass B. kinetic energy
C. weight D. potential energy

Questions 15-18.

DIRECTIONS: In answering Questions 15 to 18, use the graphs below.

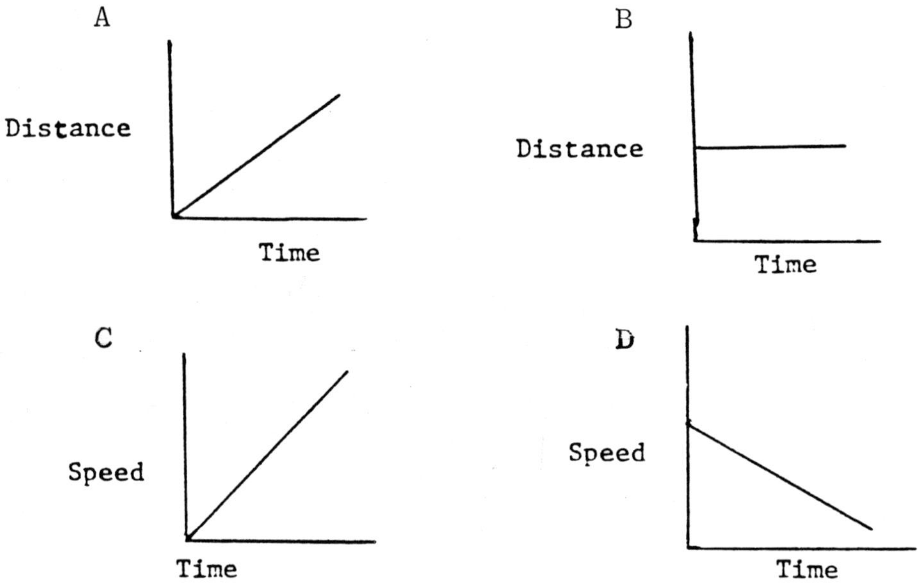

15. Which graph represents a car accelerating? 15.____

16. Which graph represents a car standing still? 16.____

17. Which graph represents a car moving at a constant speed? 17.____

18. Which graph shows a car slowing down (decelerating)? 18.____

19. Three scientific laws which explain the motion of objects were stated by 19.____
A. Aristotle B. Dalton C. Galileo D. Newton

20. The tendency of an object to keep moving is 20.____
A. inertia B. speed C. acceleration D. force

21. Centipedal force acts in a direction 21.____
A. along the path of motion
B. away from the center of a circle
C. toward the center of a circle
D. none of the above

22. _____ are types of matter which can occupy the same space at the same time. 22.___
   A. Gas and solid B. Liquid to gas
   C. Gas and liquid D. None of the above

23. The boiling point is the temperature at which a substance changes from 23.___
   A. solid to liquid B. liquid to gas
   C. solid to gas D. liquid to solid

24. A physical property of a substance is that it 24.___
   A. has a red color
   B. reacts slowly when mixed with an acid
   C. burns in oxygen
   D. explodes when heated

25. To find the density of a substance, you MUST know its mass and 25.___
   A. volume B. weight C. composition D. shape

26. Isotopes are atoms with the same 26.___
   A. number of protons B. mass number
   C. number of neutrons D. atomic mass

27. The energy levels within an atom contain 27.___
   A. protons B. neutrons C. isotopes D. electrons

28. Al, Cu, and O are symbols for 28.___
   A. compounds B. elements C. mixtures D. gases

29. The elements on a periodic table are arranged according to increasing 29.___
   A. atomic mass B. stability
   C. activity D. atomic number

30. The mass of an atom depends on the number of _____ and _____. 30.___
   A. protons; neutrons B. protons; electrons
   C. electrons; neutrons D. electrons; isotopes

Questions 31-34.

DIRECTIONS: Match the following terms.

31. Noble gases 31.___

32. Ion 32.___

33. Halogens 33.___

34. Alkalis 34.___

A. An atom or molecule that has an electric charge

B. An electron arrangement of completely filled shells

C. The most reactive non metals

D. The way an atom reacts with other atoms

E. Family of most reactive metals

35. What form of matter may be separated by chemical but NOT by physical means? 35.___
A. Element B. Compound C. Solution D. Mixture

36. Bonding by electron transfer is ____ bonding. 36.___
A. electronic B. shared C. covalent D. ionic

37. The particle that results from a covalent bond is a(n) 37.___
A. ion B. molecule C. electron D. isotope

38. The MOST widely used metal is 38.___
A. aluminum B. iron C. copper D. tin

39. Silver and gold are members of the ____ family. 39.___
A. copper B. nickel C. zinc D. platinum

40. The MOST abundant metal in the earth's crust is 40.___
A. aluminum B. iron C. carbon D. sodium

Questions 41-45.

DIRECTIONS: Match the following elements to the proper family. Answers may be used more than once as needed.

41.	Neon	A. Halogen family	41.___
42.	Sulfur	B. Nitrogen family	42.___
43.	Satatine	C. Noble gases	43.___
44.	Chlorine	D. Oxygen family	44.___
45.	Phosphorous		45.___

KEY (CORRECT ANSWERS)

1. B	11. D	21. C	31. B	41. C
2. D	12. C	22. D	32. A	42. D
3. D	13. D	23. B	33. C	43. A
4. A	14. B	24. A	34. E	44. A
5. C	15. C	25. A	35. B	45. B
6. B	16. B	26. A	36. D	
7. D	17. A	27. D	37. B	
8. C	18. D	28. B	38. B	
9. A	19. D	29. D	39. A	
10. B	20. A	30. A	40. A	

TEST 2

DIRECTIONS: Each question or incomplete statement is followed by several suggested answers or completions. Select the one that BEST answers the question or completes the statement. *PRINT THE LETTER OF THE CORRECT ANSWER IN THE SPACE AT THE RIGHT.*

1. _____ electrons are in the outer energy level of each halogen. 1.___
   A. One B. Two C. Seven D. Eight

2. The MOST abundant halogen is 2.___
   A. fluorine B. chlorine C. bromine D. iodine

3. Transistors are made using the element 3.___
   A. silicon B. carbon C. tin D. lead

4. Graphite and diamonds are _____ of carbon. 4.___
   A. isotopes B. allotropes C. isomers D. compounds

5. The ratio between hydrogen and oxygen is two to one in 5.___
   A. fats B. carbohydrates C. proteins D. vitamins

6. All of these are organic compounds EXCEPT 6.___
   A. vitamins B. soap
   C. table sugar D. table salt

7. When petroleum, a fossil fuel, undergoes fractional distillation, _____ is obtained in one of the fractions. 7.___
   A. alcohol B. gasoline C. coal D. natural gas

8. _____ provide(s) the GREATEST source of energy per gram for the body. 8.___
   A. Fats B. Proteins C. Sugars D. Starches

9. $H_2 + Cl_2 \rightarrow 2HCl$ is a chemical 9.___
   A. symbol B. equation C. formula D. decomposition

10. Which of these equations is balanced? 10.___
    A. $2H_2 + O_2 \rightarrow H_2O$ B. $Na + Cl_2 \rightarrow 2NaCl$
    C. $N_2 + 3H_2 \rightarrow 2NH_3$ D. $2C + O_2 \rightarrow 2CO_2$

11. Which term BEST describes the energy changes during the burning of magnesium? 11.___
    A. Exothermic B. Slow
    C. Endothermic D. Electrolytic

12. The mass of all the substances before a chemical reaction is _____ the mass of all the substances after the reaction. 12.___
    A. more than B. less than C. the same as

13. The rate at which vibrations are produced per unit of time is called 13.____
A. amplitude B. intensity C. frequency D. pitch

14. Sound waves of frequencies greater than 20,000 cycles/second are called 14.____
A. overtones B. reverberations
C. echos D. ultrasonic sounds

15. The intensity of a sound is measured in units called 15.____
A. hertz B. decibels
C. fundamentals D. none of the above

16. The measure of the MAXIMUM movement of a wave from its normal rest position is called the wave's 16.____
A. amplitude B. intensity C. frequency D. pitch

17. The organization of the notes in time is 17.____
A. rhythm B. timbre
C. resonance D. natural frequency

18. The image produced by a plane mirror is 18.____
A. real B. inverted C. enlarged D. virtual

19. With all mirrors, the angle of reflection equals the angle of 19.____
A. incidence B. refraction
C. diffraction D. polarization

20. When you look at a pencil in a glass of water, it appears bent because the light is 20.____
A. dispersed B. defracted C. refracted D. polarized

21. A book looks blue to you because blue light is 21.____
A. absorbed B. reflected C. refracted D. transmitted

22. The image produced by a convex lens may be ____ the object. 22.____
A. smaller than B. larger than
C. the same size as D. all of the above

23. ____ images are formed by a convex lens. 23.____
A. Real B. Virtual
C. Both real and virtual

24. The eyepiece of a compound microscope *usually* magnifies about 24.____
A. 3x B. 5x C. 10x D. 40x

25. The concave lens can be used to correct 25.____
A. farsightedness B. nearsightedness
C. color blindness D. astigmatism

26. A(n) ____ wave is a sound wave. 26.____
A. transverse B. compressional
C. electromagnetic

27. Sound waves travel FASTEST through a 27.____
A. vacuum B. gas C. liquid D. solid

28. The loudness or volume of a sound is determined by the sound's 28.____
A. amplitude B. frequency C. pitch D. wavelength

29. A _____ produces sounds by means of a vibrating reed. 29.____
A. piano B. trumpet C. violin D. clarinet

30. A physical property of a substance is its 30.____
A. heat B. pressure
C. specific heat capacity D. all of the above

31. The SI unit used to express heat energy is the 31.____
A. calorie B. joule C. pascal D. degree

32. Force per unit area is an expression of 32.____
A. work B. energy C. pressure D. power

33. The relationship between the temperature and the volume of a gas at constant pressure is stated in _____ law. 33.____
A. Pascal's B. Boyle's C. Charles' D. Joule's

34. Which temperature-time graph BEST represents heating a mixture of ice and water while the ice is still melting? 34.____

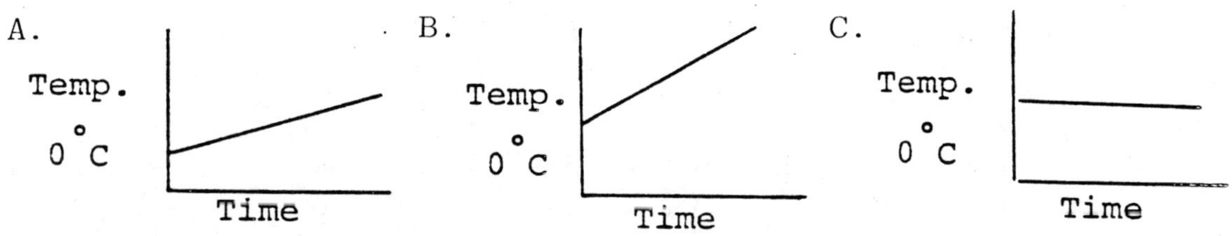

D. None of the above

35. What method of heat movement occurs BEST through metallic solids? 35.____
A. Radiation B. Convection C. Reflection D. Conduction

36. Like electric charges 36.____
A. repel B. neutralize C. attract D. cancel

37. A neutral object will become positively charged if it loses 37.____
A. ions B. neutrons C. protons D. electrons

38. _____ is NOT a good electrical insulator. 38.____
A. Plastic B. Aluminum C. Rubber D. Glass

39. When one bulb in a string of Christmas lights is removed, all of the lights go out. 39.____
The lights are connected
A. in series B. in parallel

40. The poles of two magnets are near each other and repel. The poles could NOT be 40.___
   A. two north poles   B. a north and a south pole
   C. two south poles   D. both A and C above

KEY (CORRECT ANSWERS)

1. C	11. A	21. B	31. B
2. B	12. C	22. D	32. C
3. A	13. C	23. C	33. C
4. B	14. D	24. C	34. C
5. B	15. B	25. B	35. D
6. D	16. A	26. B	36. A
7. B	17. A	27. D	37. D
8. A	18. D	28. A	38. B
9. B	19. A	29. D	39. A
10. C	20. C	30. C	40. B

EXAMINATION SECTION
TEST 1

DIRECTIONS: Each question or incomplete statement is followed by several suggested answers or completions. Select the one that BEST answers the question or completes the statement. *PRINT THE LETTER OF THE CORRECT ANSWER IN THE SPACE AT THE RIGHT.*

Questions 1-10.

DIRECTIONS: Questions 1 through 10 are of the matching type. Column I lists units which can be properly matched with the physical terms given in Column II. For the items of Column I, write in the space at the right the letter in front of the option in Column II to which it pertains.

COLUMN I	COLUMN II	
1. ampere	A. capacitance	1.___
2. calorie	B. current	2.___
3. centimeter	C. force	3.___
4. coulomb	D. heat	4.___
5. dyne	E. impedance	5.___
6. henry	F. inductance	6.___
7. joule	G. magnetic field	7.___
8. ohm	H. potential	8.___
9. volt	I. power	9.___
10. watt	J. pressure	10.___
	K. quantity of charge	
	L. reluctance	
	M. work	

11. The term *fusion* means MOST NEARLY 11.___
    A. boiling B. dissolving
    C. melting D. sublimating

12. Of the following, the prefix which means one hundredth in the metric system is 12.____
A. centi- B. deci- C. hecto- D. kilo-

13. The one of the following which is NOT a measure of length is the 13.____
A. cm B. m C. ml D. mm

14. Ordinary body temperature is about 37° on the _____ scale. 14.____
A. absolute
B. Baume
C. centigrade
D. Fahrenheit

15. The one of the following relations that is MOST valid is 15.____
A. 1 cm = .39 inches
B. 1 H.P. = 550 foot-pounds
C. 1 lb. = 2.2 kg.
D. 1 meter = 100 inches

16. Of the following, the one that is NOT a *vector* quantity is 16.____
A. angular momentum
B. frictional force
C. mass
D. velocity

17. Of the following, the one that is a vector quantity is 17.____
A. capacity
B. charge
C. electric field
D. potential

18. The arithmetic mean of 1/3 and 1/6 is 18.____
A. 1/18 B. 1/12 C. 1/4 D. 1/2

19. The number 2380 may also be written as 2.38 x 19.____
A. 10^2 B. 10^3 C. 10^4 D. 10^5

20. 10^{-2} is equal to 20.____
A. 0.01 B. 0.1 C. 1 D. 100

21. Density is GENERALLY defined as the ratio of 21.____
A. mass to volume
B. mass to weight
C. volume to mass
D. weight to mass

22. The capacity to do work is termed 22.____
A. efficiency
B. energy
C. mechanical advantage
D. velocity

23. A is a heavy body; B is a lighter body. 23.____
From this information, we can conclude that
A. A has a greater volume than B
B. A has a larger amount of inertia than B
C. if A and B are dropped together, A will strike the ground first
D. the same force will accelerate A faster than B

24. A 10 lb. weight is suspended on a spring balance in an elevator. 24.____
If the elevator is descending with a constant speed of 32 ft/sec, the spring balance will read _____ lbs.
A. 0 B. 10 C. 20 D. 32

25. A body moving in a circle at constant speed is 25.___
   A. constantly being accelerated along the tangent to the circle at any point
   B. constantly being accelerated away from the center of the circle
   C. constantly being accelerated toward the center of the circle
   D. not being accelerated at all

26. A tuning fork has a frequency of 300 c.p.s. in air where the velocity of sound is 1100 ft/sec. 26.___
The wavelength of the sound wave in air is _____ ft.
   A. 0.3 B. 3.0 C. 3.3 D. 33.0

27. The mercurial barometer reading at sea level is GENERALLY about _____ inches. 27.___
   A. 30 B. 50 C. 76 D. 100

28. The anaeroid is a type of 28.___
   A. ammeter B. barometer
   C. thermometer D. voltmeter

29. Pressure differences are GENERALLY determined by the use of a 29.___
   A. calorimeter B. colorimeter
   C. hydrometer D. manometer

30. A man wants to build a barometer but the tube he wants to use has a diameter twice as large as the diameter of the standard barometer he is using as a model. 30.___
In order to get the correct pressure reading, he will have to multiply the scale reading by
   A. 1/2 B. 1 C. 2 D. 4

KEY (CORRECT ANSWERS)

1. B	11. C	21. A
2. D	12. A	22. B
3. A	13. C	23. B
4. K	14. C	24. B
5. C	15. A	25. C
6. F	16. C	26. C
7. M	17. C	27. A
8. E	18. C	28. B
9. H	19. B	29. D
10. I	20. A	30. B

TEST 2

DIRECTIONS: Each question or incomplete statement is followed by several suggested answers or completions. Select the one that BEST answers the question or completes the statement. *PRINT THE LETTER OF THE CORRECT ANSWER IN THE SPACE AT THE RIGHT.*

1. If one is most comfortable when the temperature is 68°F, which of the following temperatures would be the MOST agreeable? 1.____
   A. 8°C
   B. 40°C
   C. 232° absolute
   D. 293° absolute

2. Of the following, the one that has the HIGHEST specific heat is 2.____
   A. glass
   B. iron
   C. lead
   D. water

3. The Leyden jar is a type of 3.____
   A. dry cell
   B. electrical condenser
   C. electroscope
   D. rheostat

4. The farad is a unit of 4.____
   A. capacitance
   B. inductance
   C. reluctance
   D. resistance

5. A step-down transformer is used to change 5.____
   A. alternating current to direct current
   B. direct current to alternating current
   C. high voltage to low voltage
   D. low voltage to high voltage

6. Electrical energy is transformed into mechanical energy by means of a 6.____
   A. generator
   B. motor
   C. storage cell
   D. transformer

7. The voltage supplied to each of five lamps arranged in parallel in a 550-volt circuit is _____ volts. 7.____
   A. 5
   B. 110
   C. 550
   D. 2750

8. In the electrolysis of water, oxygen collects at the 8.____
   A. anode
   B. cathode
   C. electrode
   D. node

9. A spoon to be silverplated should be 9.____
   A. coated with a non-conductor
   B. hung at the anode
   C. hung at the cathode
   D. hung between the anode and the cathode

10. Of the following, the POOREST conductor of electricity is 10.___
A. brass B. copper C. lead D. silver

11. The resistance of two resistors in parallel is 11.___
A. greater than either of them
B. greater than the smaller resistor
C. smaller than either of them
D. the geometric mean of the two resistances

12. The force between a magnetic pole of strength m and an electric charge of strength Q which are both at rest at a distance r cm apart is 12.___
A. $\frac{m + Q}{r}$ B. mQ/r C. mQ/r^2 D. 0

13. A long bar of soft iron placed at an angle with respect to a uniform magnetic field tends to _____ the field. 13.___
A. move in the direction of
B. move opposite to the direction of
C. turn until its long axis is parallel to
D. turn until its long axis is perpendicular to

14. Rectifiers are USUALLY 14.___
A. diodes B. pentodes
C. thyratrons D. triodes

15. The controlling element in a triode is the 15.___
A. anode B. cathode C. filament D. grid

16. The formation of sharp shadows by a point source of light furnishes experimental evidence of the 16.___
A. finite speed of light
B. inverse square law
C. principle of reversibility of light
D. rectilinear propagation of light

17. The speed of propagation of x-rays in a vacuum is 17.___
A. greater than the speed of light
B. less than the speed of light
C. the same as the speed of light
D. unknown

18. Three of the following are properly associated with one another.
The fourth that is NOT properly associated with the others is 18.___
A. beta ray B. gamma ray C. radio wave D. x-ray

19. Raising the voltage across an x-ray tube results in x-rays of 19.___
A. greater speed
B. higher frequencies
C. longer wavelengths
D. the same wavelength but greater energy

20. Of the following, the LONGEST wavelength is possessed by 20.___
A. infra-red rays B. ultraviolet rays
C. visible violet rays D. x-rays

21. Of the following, the one that is NOT an elementary particle is the 21.___
A. betatron B. electron C. neutron D. positron

22. Isotopes are atoms that have _____ atomic weight(s) and(but) _____ atomic number(s). 22.___
A. different; different B. different; the same
C. the same; the same D. the same; different

23. An electron volt is a unit of 23.___
A. charge B. energy C. power D. voltage

24. Radium was discovered by 24.___
A. Curie B. Davy
C. Einstein D. Metchnikoff

25. If the activity of I^{131} decreases to $\frac{1}{2}$ in 8 days, in another 8 days its activity compared with that at the start will be 25.___
A. 1/16 B. 1/8 C. 1/4 D. 0

26. *Hypo* is used in photography as a 26.___
A. developer B. fixer
C. reducer D. sensitizer

27. A test meter is NOT usually used to measure 27.___
A. capacitance B. current
C. resistance D. voltage

28. Speed independent of load is BEST obtained from a(n) _____ motor. 28.___
A. series B. shunt
C. synchronous D. induction

29. A metal which can be drawn into fine threads is called 29.___
A. amorphous B. ductile
C. elastic D. flexible

30. The alloy composed of tin and lead is known as 30.___
A. brass B. bronze
C. monel metal D. solder

KEY (CORRECT ANSWERS)

1. D	11. C	21. A
2. D	12. D	22. B
3. B	13. C	23. B
4. A	14. A	24. A
5. C	15. D	25. C
6. B	16. D	26. B
7. C	17. C	27. A
8. A	18. A	28. C
9. C	19. B	29. B
10. C	20. A	30. D

TEST 3

DIRECTIONS: Each question or incomplete statement is followed by several suggested answers or completions. Select the one that BEST answers the question or completes the statement. *PRINT THE LETTER OF THE CORRECT ANSWER IN THE SPACE AT THE RIGHT.*

1. A 150 pound body can just float in fresh water. The volume of this body is APPROXIMATELY _____ ft^3. 1.____
   A. 0.15 B. 1.5 C. 2.5 D. 15

2. A radian is APPROXIMATELY 2.____
   A. 37° B. 57° C. 77° D. 97°

3. The vector sum of two forces, one 3 pounds and the other 4 pounds, 3.____
   A. is 1 pound
   B. is 5 pounds
   C. is 7 pounds
   D. cannot be determined from the information given

4. When bodies initially at rest receive uniform acceleration, the distance traversed increases as the 4.____
   A. change in momentum B. instantaneous velocity
   C. square of the time D. square root of the time

5. Simple harmonic motion is produced by a restering force which is 5.____
   A. constant in magnitude but not in direction
   B. proportional to the velocity
   C. proportional to the displacement
   D. varying in magnitude and constant in direction

6. The resulting total momentum of the fragments of a shell just after it bursts in flight as compared to the momentum of the shell just before it bursts 6.____
   A. is converted to kinetic energy of the fragments
   B. is decreased
   C. is increased
   D. remains unchanged

7. Young's modulus of elasticity varies with the 7.____
   A. applied force
   B. area of the material being tested
   C. kind of material tested
   D. length of the material being tested

8. A book is resting on a table. 8.___
According to Newton's Third Law, the force which is the reaction to the gravitational force on the book is
A. applied by the book at its center of gravity
B. the force exerted by the book on the earth
C. the force exerted by the book on the table
D. the force exerted by the table on the book

9. Disregarding friction, which of the quantities listed below is unchanged when a body is falling freely through the air? 9.___
A. Kinetic energy
B. Potential energy
C. The sum of the kinetic energy and potential energy
D. The difference between the kinetic energy and potential energy

10. An organ pipe closed at one end emits a fundamental frequency of 600 c.p.s. 10.___
It can also emit a sound the frequency of which is ______ c.p.s.
A. 300 B. 1200 C. 1800 D. 2400

11. When a fluid flows through a constriction in a horizontal pipe, the ______ increases. 11.___
A. density B. potential energy
C. pressure D. velocity

12. In the steady state, the rate at which heat is conducted through a wall does NOT depend on the 12.___
A. specific heat of the wall
B. temperature gradient
C. thermal conductivity of the wall
D. wall thickness

13. The door of a refrigerator is left open. 13.___
If the refrigerator is operating properly, the room temperature will
A. decrease
B. increase
C. increase or decrease, depending on whether it is a gas or electric refrigerator
D. not change

14. Molecules are dissociated in aqueous solution for the following reason: 14.___
A. Cosmic radiation
B. The electrostatic forces holding the atoms together are weakened in a medium of high dielectric constant
C. The few ions always present in water collide with the molecules breaking them apart
D. The molecules of the water bombard the molecules of the solute more vigorously than air at the same temperature

15. The capacity of a parallel plate condenser of area A and plate separation d is proportional to 15.___

A. $\frac{A}{d}$ B. $\frac{A}{d^2}$ C. dA D. $\frac{d^2}{A}$

16. Three charges +20 e.s.u., +20 e.s.u., and -20 e.s.u., respectively, are at the corners of an equilateral triangle 10 cm on a side. 16.___
The work required to move the negative charge to a point very far away is _____ ergs.

A. 0 B. 4 C. 40 D. 80

17. Two points in a circuit differ in potential by one volt if it requires _____ to move 1 coulomb between the points. 17.___

A. 1 erg B. 1 joule C. 1 watt D. 10 watts

18. A motor draws 2 amperes from a 110-volt circuit while driving a small lathe. Heat losses in its windings amount to 20 watts. 18.___
Its back E.M.F. is _____ volts.

A. 10 B. 55 C. 100 D. 110

19. A potentiometer is a useful instrument for measuring E.M.F. because it 19.___

A. draws no current from the source being measured
B. has a high internal resistance
C. has a low internal resistance
D. measures current as well as voltage

20. The turns ratio of a transformer is 1 to 10. A source of 100 volt, 60 c.p.s. is connected to the primary. The open circuit voltage of the secondary will be 20.___

A. 600 B. 1,000 C. 3,600 D. 10,000

21. To determine the power dissipation in a heater, one may connect an ammeter _____ with the heater. 21.___

A. and voltmeter in parallel
B. and voltmeter in series
C. in parallel and a voltmeter in series
D. in series and a voltmeter in parallel

22. An electromagnetic wave with a frequency of about 100 kc/s 22.___

A. has a wavelength about 2 feet
B. has a wavelength about 3000 meters
C. has a wavelength in the visible region
D. may have any wavelength

23. The force which can be exerted by an electromagnet will NOT be lowered if the 23.___

A. current is reduced
B. iron core is removed
C. number of windings is doubled and the current halved
D. number of windings is halved and the current kept the same

24. The precision of an ordinary shop volt - ohm - milliammeter is USUALLY _____% of full scale. 24.___
A. 0.5 B. 1 C. 5 D. 25

25. The center tap of a filament transformer is USUALLY connected to the 25.___
A. cathode B. grid C. plate D. screen

26. Hum pick-up due to the heater leads is reduced by 26.___
A. twisting the leads
B. using thicker wires
C. using thinner wires
D. using separate filament transformers for low level stages

27. To make an absolute determination of the wavelength of light, one should 27.___
A. measure the velocity of light
B. use a diffraction grating
C. use a prism spectrometer
D. use a refractometer

28. The formation of a mirage is due to 28.___
A. diffraction B. interference
C. reflection D. refraction

29. Light source I has four times the candlepower of source II. 29.___
In order that the two sources provide equal illumination on a fixed screen normal to the incident light, source II should be _____ as far from the screen as source I.
A. 1/4 B. 1/2 C. twice D. 4 times

30. If a simple converging lens clearly shows chromatic aberration, 30.___
A. it forms a poor image when monochromatic light is used
B. its focal length for blue light is less than for red light
C. its focal length for red light is less than for blue light
D. the rays that pass through it obliquely from an object point off the axis do not converge through a single point

KEY (CORRECT ANSWERS)

1. C	11. D	21. D
2. B	12. A	22. B
3. D	13. B	23. C
4. C	14. B	24. C
5. C	15. A	25. A
6. D	16. D	26. A
7. C	17. B	27. B
8. B	18. C	28. D
9. C	19. A	29. B
10. C	20. B	30. B

TEST 4

DIRECTIONS: Each question or incomplete statement is followed by several suggested answers or completions. Select the one that BEST answers the question or completes the statement. *PRINT THE LETTER OF THE CORRECT ANSWER IN THE SPACE AT THE RIGHT.*

1. In order to increase the magnifying power of a compound microscope, one should select an objective lens of 1.___
   A. larger diameter B. larger focal length
   C. shorter focal length D. smaller diameter

2. The group of light characteristics which can be accounted for equally well on the assumption that light is either wave motion or a simple corpuscular phenomenon is 2.___
   A. aberration, color, polarization
   B. interference, refraction, diffraction
   C. rectilinear propagation, intensity, reflection
   D. speed, intensity, scattering, photoelectric effect

3. When light falls on a metal surface and electrons are emitted, the energy of the fastest ones emitted increases as the _____ of the light increases. 3.___
   A. frequency B. intensity C. velocity D. wavelength

4. When the nucleus of a radioactive element emits a beta particle, its atomic _____ by one. 4.___
   A. number is decreased B. number is increased
   C. weight is decreased D. weight is increased

5. Emission of a gamma ray 5.___
   A. changes the charge on the nucleus
   B. changes the mass number of the nucleus
   C. has no effect on the chemical nature of the isotope
   D. is called transmutation

6. Pyrex glass is used in laboratories because of its 6.___
   A. large expansion coefficient
   B. large specific heat
   C. small expansion coefficient
   D. small specific heat

7. After which of the following accidents in the dark room should one of the processing solutions be discarded? 7.___
   A. Developer gets into fixer
   B. Fixer gets into developer
   C. Fixer gets into stop bath
   D. Stop bath gets into fixer

8. Sulfuric and chromic acid cleaning mixture can GENERALLY be used until its color becomes 8.___
   A. blue B. green C. red D. violet

9. A power supply delivers 300 volts. 9.___
   The connection that should be installed as the outlet on the chassis is a
   A. set of alligator clips B. set of binding posts
   C. female socket D. male socket

10. Of the following, the LEAST satisfactory material used in a high vacuum system is 10.___
   A. grease B. plastic C. wax D. wood

11. Sudden heating or chilling is BEST tolerated by 11.___
   A. monex B. pyrex C. quartz D. soft glass

12. Good electrical insulators may NOT be made from 12.___
   A. polyethylene B. polystyrene
   C. sulfur D. wood

13. Telescope mirrors are USUALLY made by 13.___
   A. grinding two pieces of glass together
   B. grinding with a special emery wheel
   C. grinding with diamond dust
   D. using a glassblower's lathe

14. Glass blowing with pyrex glass is USUALLY done with a(n) _____ torch. 14.___
   A. gas-air B. gas-oxygen
   C. hydrogen-oxygen D. acetylene

15. A diffusion pump must ALWAYS be 15.___
   A. operated with a fore pump
   B. operated with a liquid air trap
   C. operated with an ionization gauge
   D. water cooled

16. Highest capacitances are USUALLY 16.___
   A. air condensers B. electrolytic condensers
   C. mica condensers D. oil filled

17. Satisfactory glass to metal seals may NOT be made by 17.___
   A. cementing
   B. machining of metal to fit the glass
   C. rubber gaskets
   D. soldering

18. Glass may NOT be made conducting by 18.___
   A. painting it with colloidal graphite
   B. evaporating a metallic film onto it
   C. incorporating metallic powders into it
   D. sputtering a metallic film onto it

19. A follower rest is used in 19.___
A. cutting intricate shapes on a jig saw
B. cutting large gears
C. drilling small parts
D. turning long shafts

20. A dog is used in a lathe when the work is held 20.___
A. between centers B. in a draw-in collet
C. in a steady rest D. in an independent chuck

21. A fly cutter is USUALLY used in a 21.___
A. drill press B. lathe
C. milling machine D. shaper

22. A block of brass is provided with a hole. 22.___
A close-fitting plunger will slide MOST freely in the hole if the plunger is made of
A. brass B. copper C. lead D. steel

23. Silver soldering 23.___
A. gives a stronger joint than soft soldering
B. gives a weaker joint than soft soldering
C. may be used to bond aluminum to iron
D. requires lower temperatures than soft soldering

24. A round rod $\frac{1}{4}$" diameter will have a good sliding fit in a hole having a diameter of _____ thousandths of an inch. 24.___
A. 250 B. 250.1 C. 253 D. 260

25. A lathe is NOT generally used to 25.___
A. cut gears B. cut screws
C. cut tapers D. plane surfaces

26. The diameter size of a tap drill is _____ that of a body drill. 26.___
A. 10% larger than B. smaller than
C. the same as D. 20% larger than

27. In soldering, flux is used to eliminate 27.___
A. combustion B. dirt
C. oxidation D. reduction

28. The limit of micrometer precision is 28.___
A. .0001 inch B. .01 inch
C. .1 micron D. 10 mils

29. The one of the following drills NEAREST to the correct size of body drill for a No. 6 screw is the 29.___
A. 1/4" B. No. 4 C. No. 27 D. No. 58

30. The purpose of case hardening of steel rod is to 30.___
A. harden the core B. harden the surface
C. make it brittle D. prepare it for polishing

KEY (CORRECT ANSWERS)

1. C	11. C	21. A
2. C	12. D	22. D
3. A	13. A	23. A
4. B	14. B	24. C
5. C	15. A	25. A
6. C	16. B	26. B
7. B	17. B	27. C
8. B	18. C	28. A
9. C	19. D	29. C
10. D	20. A	30. B

EXAMINATION SECTION

TEST 1

DIRECTIONS: Each question or incomplete statement is followed by several suggested answers or completions. Select the one that BEST answers the question or completes the statement. *PRINT THE LETTER OF THE CORRECT ANSWER IN THE SPACE AT THE RIGHT.*

1. The SMALLEST unit of a compound which possesses all the chemical characteristics of the compound is a(n) 1.___
   A. ion B. atom C. element D. molecule

2. It takes APPROXIMATELY _____ for the moon to revolve around the earth. 2.___
   A. twenty-four hours
   B. one week
   C. four weeks
   D. three months

3. The atomic weight of an element is APPROXIMATELY equal to the number of _____ the nucleus. 3.___
   A. protons in
   B. neutrons in
   C. electrons surrounding
   D. protons and neutrons in

4. A _____ displays kinetic energy. 4.___
   A. waterfall
   B. parked car
   C. storage battery
   D. tightly wound clock spring

5. The fact that the chemical compound ethane has the formula C_2H_6 means that one molecule of this compound contains _____ atoms of _____. 5.___
   A. two; carbon
   B. six; carbon
   C. six; ethane
   D. eight; ethane

6. What is the MOST abundant gas in the earth's atmosphere? 6.___
   A. Oxygen
   B. Hydrogen
   C. Nitrogen
   D. Carbon dioxide

7. Light beams of different colors have DIFFERENT 7.___
   A. opacities
   B. velocities
   C. intensities
   D. frequencies

8. Why does air over a hot surface rise? 8.___
   It(s)
   A. molecules become heavier
   B. molecules become lighter
   C. is more dense than surrounding air
   D. is less dense than surrounding air

9. What kind of heat transfer is particularly effective for liquids and gases? 9.___
   A. Radiation
   B. Convection
   C. Conduction
   D. Irradiation

10. Potential energy may be thought of as the energy of 10.____
A. speed B. power C. motion D. position

11. An increase in the barometric pressure is USUALLY a forecast of 11.____
A. rain
B. high winds
C. clearing skies
D. a sudden rise in temperature

12. Which of the following properties of hydrogen is a chemical property? 12.____
It
A. is a gas
B. is colorless
C. burns in oxygen
D. liquifies at -253°C

13. The difference between the freezing and boiling points of pure water at sea level is _____ degrees. 13.____
A. 100 centigrade
B. 180 centigrade
C. 212 Fahrenheit
D. 273 Fahrenheit

14. The X-ray was originally discovered by 14.____
A. Roentgen B. Barium C. Wheeler D. Marconi

15. The part of a telephone which changes your voice's sound waves into electrical current is called a 15.____
A. receiver
B. electromagnet
C. transmitter
D. diaphragm

16. The sun is a mass of 16.____
A. liquids
B. solids
C. gases
D. liquids and solids

17. Heat is received from the sun by 17.____
A. radiation
B. a convection current
C. conduction
D. all of the above

18. Of the following, _____ is not necessary to start a fire. 18.____
A. fuel B. oxygen C. heat D. a match

19. The spinal cord in humans controls 19.____
A. reflex actions
B. breathing
C. heart beat
D. conscious action

20. In plants, the fruit is the 20.____
A. flower B. seed C. ovary D. stigma

21. Rickets is caused by a lack of vitamin 21.____
A. A B. C C. D D. B complex

22. Of the following, _____ is not a necessary condition for life. 22.____
A. enough oxygen and pressure
B. enough water
C. proper food and temperature
D. all are necessary

23. Sound travels 23.___
  A. the same speed in all substances
  B. different speeds through different substances
  C. faster in air than in water
  D. faster in air than in iron

24. Loudness of sounds are measured by 24.___
  A. pitch B. frequency
  C. amplitude D. resonance

25. Shortening the length of a violin string 25.___
  A. increases the loudness B. decreases the pitch
  C. causes overtones D. increases the pitch

KEY (CORRECT ANSWERS)

1. D
2. C
3. D
4. A
5. A

6. C
7. D
8. D
9. B
10. D

11. C
12. C
13. A
14. A
15. C

16. C
17. A
18. D
19. A
20. C

21. C
22. D
23. B
24. C
25. D

TEST 2

DIRECTIONS: Each question or incomplete statement is followed by several suggested answers or completions. Select the one that BEST answers the question or completes the statement. *PRINT THE LETTER OF THE CORRECT ANSWER IN THE SPACE AT THE RIGHT.*

1. Objects through which light does NOT pass are 1.____
   A. translucent B. transparent
   C. transverse D. opaque

2. A stick placed in water looks bent because light is 2.____
   A. reflected B. refracted
   C. polarized D. transmitted

3. The part of the eye that can be seen and is white is the 3.____
   A. cornea B. iris C. lens D. retina

4. An electromagnet used in the home is in a 4.____
   A. switch B. doorbell
   C. light bulb D. toaster

5. A chemical change would be 5.____
   A. ice melting
   B. making oxygen from water
   C. condensation of water vapor
   D. sugar dissolving in tea

6. A substance that is a compound is 6.____
   A. iron B. steel C. sugar D. oxygen

7. The SMALLEST part of water that still has the properties of water is a(n) 7.____
   A. atom B. electron C. molecule D. proton

8. The ability to do work is 8.____
   A. power B. force C. weight D. energy

9. The glass in MOST light bulbs is 9.____
   A. translucent B. transparent
   C. opaque D. transverse

10. Machines have an efficiency of _____ percent. 10.____
    A. 200 B. between 100 and 200
    C. 100 D. less than 100

11. Sound in which the rate of vibration is irregular is 11.____
    A. overtones B. resonance C. noise D. music

12. A can of gasoline illustrates _____ energy. 12.____
    A. kinetic B. potential C. mechanical D. heat

13. A chemical reaction which gives off heat is 13.___
A. endothermic B. exothermic
C. a physical change D. slow oxidation

14. A gas necessary for plant life is 14.___
A. oxygen B. carbon monoxide
C. hydrogen D. carbon dioxide

15. The handle of a cooking pan on a stove gets hot because of 15.___
A. conduction B. convection
C. radiation D. insulation

16. The earth's attraction for an object is 16.___
A. density B. work C. weight D. mass

17. The number of states of matter is 17.___
A. one B. two C. three D. four

18. An inorganic substance would be 18.___
A. wood B. iron C. sugar D. carbohydrate

19. Moth balls disappearing is an example of 19.___
A. evaporation B. condensation
C. vaporization D. sublimation

20. The loudness of sound depends upon the 20.___
A. amplitude B. pitch
C. frequency D. overtones

21. The pulley at the top of a flag pole is a _____ class lever. 21.___
A. first B. second
C. third D. none of the above

22. Objects at rest remain at rest because of 22.___
A. weight B. gravity C. inertia D. friction

23. To stop or start the flow of electricity, we use a 23.___
A. fuse B. switch
C. transformer D. circuit breaker

24. The SMALLEST living part of a plant or animal is a(n) 24.___
A. organ B. tissue C. molecule D. cell

25. Bacteria which are ball-shaped are 25.___
A. bacilli B. spirilli C. cocci D. virus

26. Oily rags bursting into flame are an example of 26.___
A. combustion B. spontaneous combustion
C. slow oxidation D. oxidation

Questions 27-30.

DIRECTIONS: Match item in Column I with its CORRECT answer in Column II.

COLUMN I	COLUMN II	
27. A nail touching a magnet	A. Cobalt	27.___
28. Magnetic metal	B. Magnetic field	28.___
29. Each molecule acts as a magnet	C. Magnetic theory	29.___
	D. Permanent magnet	
30. Magnetic lines of force		30.___
	E. Temporary magnet	
	F. Zinc	

KEY (CORRECT ANSWERS)

1. D	11. C	21. A
2. B	12. B	22. C
3. A	13. B	23. B
4. B	14. D	24. D
5. B	15. A	25. C
6. C	16. C	26. B
7. C	17. C	27. E
8. D	18. B	28. A
9. A	19. D	29. C
10. D	20. A	30. B

TEST 3

DIRECTIONS: Each question or incomplete statement is followed by several suggested answers or completions. Select the one that BEST answers the question or completes the statement. *PRINT THE LETTER OF THE CORRECT ANSWER IN THE SPACE AT THE RIGHT.*

1. An electrical insulator is   1.___
   A. silver   B. dry air   C. copper   D. mercury

2. A physical change would be   2.___
   A. burning wood
   B. making oxygen from water
   C. melting ice
   D. mixing soda and vinegar

3. Normal atmospheric pressure is _____ lb/sq. in.   3.___
   A. 30   B. 15   C. 5   D. 45

4. A plant, such as a tree, is a(n)   4.___
   A. perennial   B. annual   C. biennial   D. dicot

5. A unit of the rate of flow of electricity is   5.___
   A. volts   B. watts   C. ohms   D. amperes

6. The pitch of a sound is determined by the   6.___
   A. amplitude   B. frequency   C. overtones   D. resonance

7. A measure of the brightness of a light source is   7.___
   A. foot candles   B. photometers
   C. candle power   D. amperes

8. A POOR conductor of heat is   8.___
   A. steel   B. aluminum   C. silver   D. glass

9. Atmospheric pressure is measured with a   9.___
   A. thermograph   B. barometer
   C. hydrometer   D. manometer

10. Oxygen makes up _____ percent of the atmosphere.   10.___
    A. 10   B. 1   C. 80   D. 20

Questions 11-18.

DIRECTIONS: From the list of answers in Column II, select the answer which matches each item of Column I, and write the letter of the answer in the space at the right.

COLUMN I	COLUMN II	
11. Product of force and distance	A. Energy	11.___
	B. Velocity	
12. Earth's attraction for an object	C. Power	12.___
13. Capacity to do work	D. Work	13.___
14. Kilowatt hour	E. Weight	14.___
15. Distance divided by time		15.___
16. Rate of doing work		16.___
17. Force times distance		17.___
18. 550 foot pounds/sec.		18.___

Questions 19-25.

DIRECTIONS: From the list of answers in Column II, select the answer which matches each item of Column I, and write the letter of the answer in the space at the right.

COLUMN I	COLUMN II	
19. Main spring of a watch	A. Brittle	19.___
20. Object in motion tends to remain in motion	B. Ductile	20.___
	C. Elastic	
21. Balsa wood	D. Hardness	21.___
22. Cast iron	E. Inertia	22.___
23. Steel cable	F. Malleable	23.___
24. Gold	G. Porous	24.___
25. Diamond	H. Tenacious	25.___

KEY (CORRECT ANSWERS)

1. B
2. C
3. B
4. A
5. D

6. B
7. C
8. D
9. B
10. D

11. D
12. E
13. A
14. A
15. B

16. C
17. D
18. E
19. C
20. E

21. G
22. A
23. H
24. F
25. D

TEST 4

DIRECTIONS: Each question or incomplete statement is followed by several suggested answers or completions. Select the one that BEST answers the question or completes the statement. *PRINT THE LETTER OF THE CORRECT ANSWER IN THE SPACE AT THE RIGHT.*

1. A comet is — 1.____
   - A. a group of stars
   - B. a constellation
   - C. a shooting star
   - D. none of the above

2. Our universe is made up of — 2.____
   - A. ten thousand galaxies
   - B. one million galaxies
   - C. one billion galaxies
   - D. not available

3. Oxygen makes up about ____% of all air. — 3.____
   - A. 1/5
   - B. 2/3
   - C. 3/5
   - D. 1/4

4. Our bodies are composed of about _____ water. — 4.____
   - A. 1/3
   - B. 2/3
   - C. 1/5
   - D. 3/8

5. If you were to catch a fish one mile below the surface of the ocean and brought it to the surface quickly, the — 5.____
   - A. fish would remain the same size
   - B. fish would swell
   - C. fish would shrink or decrease in size
   - D. none of the above

6. Life, as we know it, has the BEST chance of being found on — 6.____
   - A. Jupiter and Saturn
   - B. Mercury and Venus
   - C. Neptune and Pluto
   - D. Venus and Mars

7. Day and night are caused by the — 7.____
   - A. sun moving westward
   - B. sun moving eastward
   - C. earth revolving around the sun
   - D. earth's rotation

8. If a peeled orange were to represent the earth, the lines dividing each section from top to bottom would be called — 8.____
   - A. latitude and meridians
   - B. latitude and longitude
   - C. longitude and meridians
   - D. none of the above

9. The meridian of the city of Greenwich, England, is — 9.____
   - A. 180° longitude
   - B. 0° latitude
   - C. 180° latitude
   - D. 0° longitude

10. The number of time zones in the United States is — 10.____
    - A. 4
    - B. 3
    - C. 5
    - D. none of the above

11. When it is 9 P.M. in New York, it is _____ in California. 11.___
   A. 12 P.M.   B. 1 A.M.   C. 5 P.M.   D. 6 P.M.

12. The star that is CLOSEST to Earth is 12.___
   A. Sirius   B. Mercury   C. the sun   D. Venus

13. The earth's surface is composed of APPROXIMATELY 13.___
   A. 70% land   B. 60% water   C. 60% land   D. 70% water

14. An instrument that records earthquakes is called a 14.___
   A. seismograph   B. barometer
   C. quake-meter   D. altimeter

15. An earthquake is caused by 15.___
   A. shock waves   B. tidal waves
   C. rock faults   D. movements of earth

16. The GREATEST destruction resulting from an earthquake occurs 16.___
   A. hundreds of miles from the site of the quake
   B. at the actual site of the quake
   C. on the opposite side of the earth
   D. in the ocean

17. An earthquake in the ocean results in a 17.___
   A. whirlpool   B. decrease in the water level
   C. tidal wave   D. typhoon

18. Wearing down the earth through weathering and erosion is the result of 18.___
   A. poor conservation
   B. the wind, sun, ice, snow, and water
   C. volcanoes
   D. rivers

19. A place where a river deposits soil is called a 19.___
   A. delta   B. tributary   C. canal   D. compound

20. The law of gravity was FIRST stated by 20.___
   A. Galileo   B. Copernicus   C. Newton   D. Ptolemy

21. The sun's effect upon tides is _____ the moon's. 21.___
   A. the same as   B. $\frac{1}{4}$ greater than
   C. less than   D. twice as great as

22. Spring tides occur 22.___
   A. once a month   B. once a year
   C. twice a year   D. twice a month

23. If one were on the moon and looked into the sky, it would be 23.___
   A. blue   B. green   C. black   D. colorless

24. An eclipse of the sun in which a ring of the sun shines out from behind the black dish of the moon is called 24.___
   A. total eclipse   B. umbra
   C. penumbra   D. annular

25. The halo around the sun in a total eclipse is known as 25.___
   A. Bailey's Beads   B. the corona
   C. a prominence   D. the umbra

26. Heated air, when compared to cold air, 26.___
   A. weighs the same   B. is heavier
   C. is lighter   D. none of the above

27. A breeze coming in from the sea is the result of the 27.___
   A. land being warmer than the sea
   B. sea being warmer than the land
   C. prevailing westerly winds
   D. none of the above

28. At night on the seashore, one could expect a breeze to 28.___
   A. come in from the sea   B. move out to the sea
   C. die down   D. none of the above

29. The LOWEST level of our atmosphere is called the 29.___
   A. ionosphere   B. stratosphere
   C. troposphere   D. none of the above

30. The average air pressure per square inch on earth is 30.___
   A. 12.7   B. 13.7   C. 14.7   D. 15.7

KEY (CORRECT ANSWERS)

1. D	11. D	21. C
2. D	12. C	22. D
3. A	13. D	23. C
4. B	14. A	24. D
5. B	15. D	25. B
6. D	16. B	26. C
7. D	17. C	27. A
8. C	18. B	28. B
9. D	19. A	29. C
10. A	20. C	30. C

EXAMINATION SECTION

TEST 1

DIRECTIONS: Each question or incomplete statement is followed by several suggested answers or completions. Select the one that BEST answers the question or completes the statement. *PRINT THE LETTER OF THE CORRECT ANSWER IN THE SPACE AT THE RIGHT.*

1. A liquid boils at 150°C. 1.___
   This temperature on the Fahrenheit scale is ____ degrees.
   A. 51.5 B. 115.5 C. 212.0 D. 238.0 E. 302.0

2. When heated, which of the following have the same rate of expansion? 2.___
   A. Solids
   B. Liquids
   C. Unconfined gases
   D. Solids and liquids
   E. None of the above

3. If 0.3 calorie of heat will raise the temperature of 1 gram of a metal 10 C. degrees, what is the specific heat of the metal? 3.___
   A. 0.003 B. 0.03 C. 0.3 D. 3.0 E. 30.0

4. When *no* heat is received from an outside source, evaporation of a liquid ALWAYS results in a(n) 4.___
   A. decrease in pressure
   B. constant temperature
   C. decrease in temperature
   D. increase in pressure
   E. increase in temperature

5. What is the relative humidity at the dew point? 5.___
   A. Zero
   B. Between 1% and 99%
   C. 100%
   D. Over 100%
   E. It may be any percent as long as the air is chilled

6. If one object is set in vibration by the vibration of another, they MUST have the same 6.___
   A. frequency
   B. amplitude
   C. mass
   D. velocity
   E. rigidity

7. The energy of a sound wave increases with its 7.___
   A. wavelength
   B. phase
   C. amplitude
   D. frequency
   E. velocity

8. All the members of an orchestra struck notes at the same time but at various musical pitches. These musical notes reached the ear of a person in the rear of the auditorium at the same time. 8.___
   This would indicate that
   A. the speed of sound waves in air is approximately 1100 feet per second

B. all the instruments produced sound waves of about the same intensity
C. sound waves travel more slowly in air than through solids
D. the speed of sound waves is independent of the pitch
E. the quality of the musical notes was improved

9. The intensity of illumination at 4 feet distance, compared with that at 1 foot distance, is 9.____
A. 1/16 B. 1/8 C. 1/4 D. 4 E. 16

10. The sun appears to rise earlier than it actually does. This is due to the earth's atmosphere causing the light waves to be 10.____
A. diffused B. reflected C. dispersed
D. polarized E. refracted

11. The phenomenon that BEST supports the hypothesis that light is a form of transverse wave motion is called 11.____
A. polarization B. refraction C. interference
D. reflection E. dispersion

12. The types of images formed by a convex lens when the object is placed at varying distances are MOST like those formed by a 12.____
A. prism B. concave mirror
C. plane mirror D. convex mirror
E. concave lens

13. When light traveling from a liquid strikes a solid, MOST of the light will be 13.____
A. reflected B. diffused
C. refracted D. dispersed
E. It is impossible to tell

14. 14.____

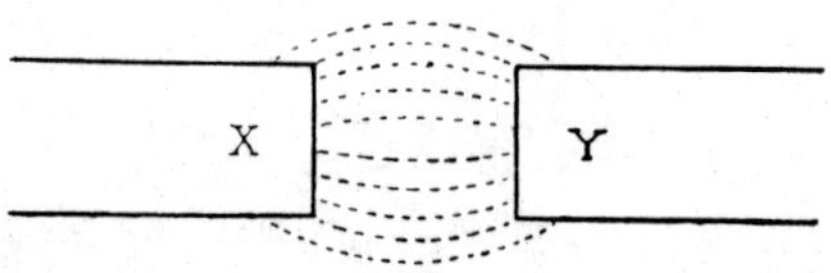

The diagram above represents a magnetic field between the ends of two magnets.
The shape of the field indicates that
A. end X is a north pole and end Y is a south pole
B. both poles are north poles
C. both poles are south poles
D. one pole is a north pole and the other pole is a south pole
E. end X is a south pole and end Y is a north pole

Questions 15-16.

DIRECTIONS: Use the information and diagram below in answering Questions 15 and 16.

A ray of monochromatic red light passes from air through three glass plates and back into the air as shown in the diagram. The index of refraction (n) for each material is given.

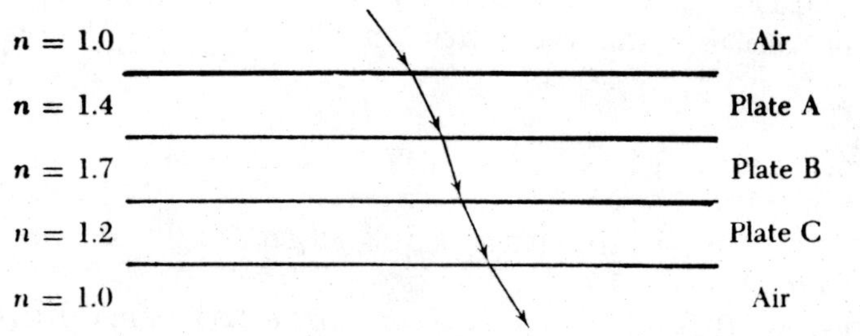

15. The frequency of the light is 15.___
   A. less in plate A than in plate C
   B. the same in air, plates A, B, and C
   C. greater in air than in plate B
   D. greater in plate A than in plate B

16. The wavelength of the light is 16.___
   A. smaller in air than in plate A
   B. smaller in air than in plate C
   C. greater in plate A than in plate C
   D. greater in plate A than in plate B

17. Observation of which of these phenomena caused questions to be raised concerning the validity of the particle theory of light? 17.___
   A. Interference
   B. Photoelectric effect
   C. Reflection from a rough surface
   D. Decrease of intensity of illumination with distance from the source

18. The phenomena of ____ BEST supports the hypothesis that light is a form of transverse wave motion. 18.___
   A. dispersion B. reflection
   C. polarization D. refraction

19. The relationship among Planck's constant h, the frequency f, and the energy E of a photon is given by 19.___
   A. $E = f/h$ B. $E = hf$ C. $f = Eh$ D. $E = h/f$

20. Two masses, M_1 and M_2, are on a horizontal, frictionless surface. M_1 has a momentum of +5.0 units, M_2 is at rest. If M_1 strikes and sticks to M_2, the change in the momentum of the system (consisting of M_1 and M_2) is ____ units. 20.___
   A. +5.0 B. +2.5 C. 0 D. -5.0

21. The discrete energy levels in an atom can BEST be explained by assuming that the electrons 21.___
   A. occupy stable positions only at distances from the nucleus where they form a standing wave
   B. obey Coulomb's Law
   C. obey the law of universal gravitation
   D. emit photons of high energy as they accelerate around the nucleus

22. A force of 2.0 newtons is applied for 4.0 seconds to a 5.0 kilogram cart in the direction of its motion. 22.___
The change in momentum of the cart is ____ kg-m./sec.
   A. 1.6   B. 8.0   C. 10   D. 20

23. Inverse proportionality is BEST illustrated by which of these relations? 23.___
   A. Length of a spring and the magnitude of a weight that stretches the spring
   B. Pressure and volume of a sample of gas kept at a constant temperature
   C. Speed of an object and the time that a constant force acts on the object in the direction of motion of the object
   D. Current and voltage in a closed circuit

24. Whether the characteristics can be observed or not, all of the following characteristics of a gas at 0°K are predicted by the ideal gas law EXCEPT the sample would 24.___
   A. occupy no volume
   B. have no electrical resistance
   C. have no thermal kinetic energy
   D. have no molecular motion

25. A man floats in water with *nearly* all of his body submerged if the 25.___
   A. average density of his body is slightly less than the density of water
   B. average density of his body is slightly greater than the density of water
   C. buoyant force on him as he floats is less than his weight
   D. mass of the displaced water is less than the mass of the man

26. The free electrons in a wire carrying electric current move at *very* high speeds V in all directions between collisions. 26.___
The overall electron movement in the direction of the electric current is at
   A. the speed of light   B. speeds higher than V
   C. the speed V   D. speeds much less than V

27. A circuit with two switches is to be wired so that a light bulb in the circuit can be turned on at either one of the switches and can be turned off at the other switch. 27.___
Which of these circuits should be used?

A.

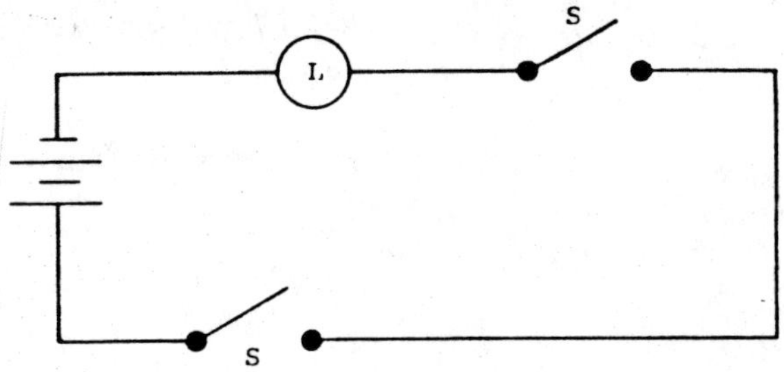

B.

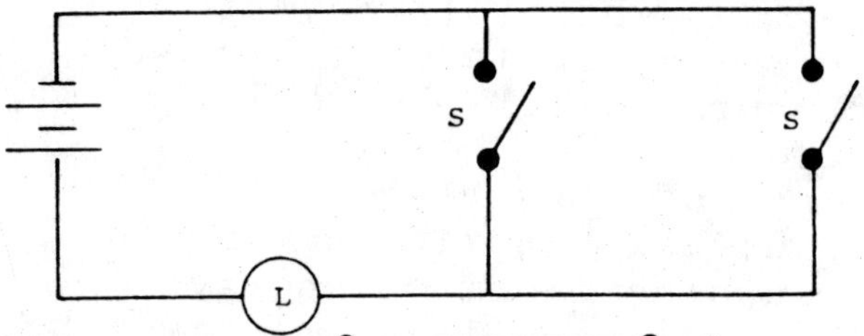

C.

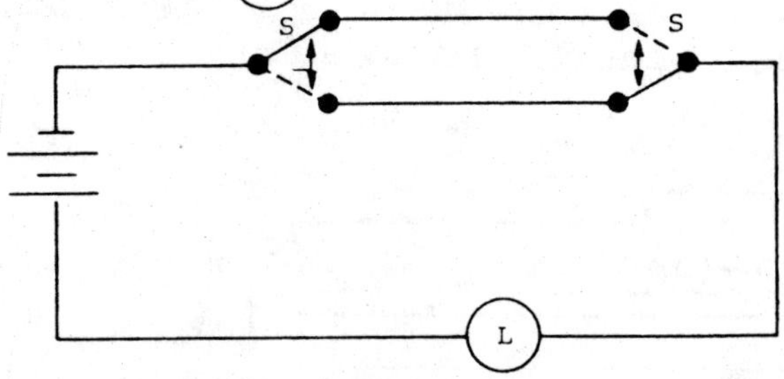

D.

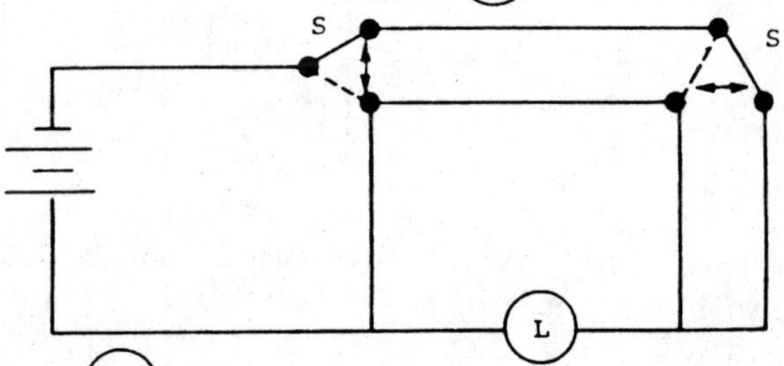

(L) Light

S —Switch

28. As an Earth satellite moves away from Earth in an elliptical orbit, which, if any, of its kinds of energy would increase? 28.___
   A. Kinetic energy *only*
   B. Potential energy *only*
   C. Both kinetic and potential energies
   D. No increase in any kind of energy

Questions 29-30.

DIRECTIONS: Questions 29 and 30 refer to the following:

A mass of 4 kilograms moves to the right at a velocity of 2 meters per second and collides with a mass of 2 kilograms which is at rest. After the collision, the 2-kilogram mass moves to the right with a velocity of 2 meters per second.

29. After the collision, the 4-kilogram mass MUST be 29.___
   A. moving to the left at 2 m/sec
   B. moving to the left at 1 m/sec
   C. stopped
   D. moving to the right at 1 m/sec

30. The data indicate that which of the following is TRUE about the kinetic energy in this collision? 30.___
   A. The kinetic energy after the collision was greater than that before.
   B. Kinetic energy was conserved.
   C. Some of the initial kinetic energy must have been converted to another form of energy.
   D. All of the initial kinetic energy must have been converted to another form of energy.

31. 31.___

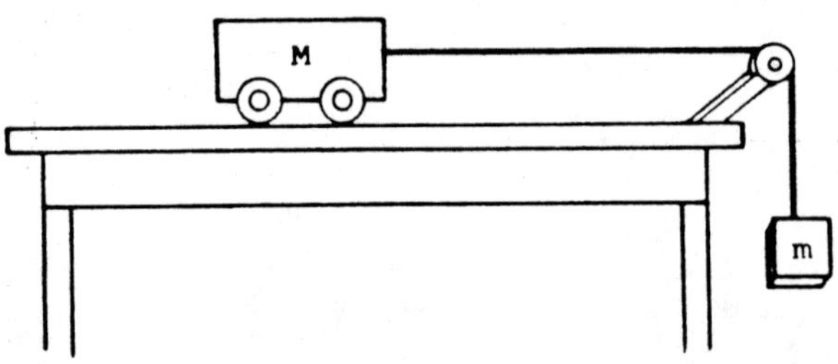

As shown above, a car of mass M is set in motion by a weight of mass m hanging over the edge of a table. There is *no* friction, and the acceleration due to gravity is g. The acceleration a of the car can be calculated from a as equal to
   A. M/mg B. mg/m
   C. mg/M+m D. Mg+mg/M+m

32. If all of these substances are at 50°F, which one will *probably* feel COLDEST to your hand? 32.___
   A. Rock
   B. Glass
   C. Iron
   D. They will all feel equally cold

33. The unequal expansion of metals makes possible the operation of 33.___
   A. hydrometers B. mercury thermometers
   C. hygrometers D. thermostats

34. A change from 68°F to 77°F is the same as ____ degrees change on the Centigrade scale. 34.___
A. 5 B. 16.2 C. 20 D. 25

35. In a vacuum, transfer of heat is possible 35.___
A. only by conduction
B. by conduction and convection
C. by convection and radiation
D. only by radiation

36. It was found that more heat was lost after a furnace pipe in the cellar was covered with a single thin sheet of asbestos paper than when the shiny bare metal was exposed to the air. 36.___
How might this be accounted for? The
A. asbestos must have been of a lighter color than the metal
B. cellar must have been at a lower temperature than the upstairs rooms
C. asbestos must have had a smoother surface than the metal
D. asbestos was a better radiator of heat than the metal

37. If water at 4°C is either heated or cooled, it will 37.___
A. release heat B. become less dense
C. contract D. solidify

38. If the mass and temperature of an enclosed gas remained constant while the volume was doubled, the density of the gas 38.___
A. was doubled B. was halved
C. decreased fourfold D. remained unchanged

Questions 39-41.

DIRECTIONS: Use the diagram and information below in answering Questions 39 to 41.

In the following circuit, disregard the internal resistance of the source, the resistance of the connecting wires, and the effects of the meters.

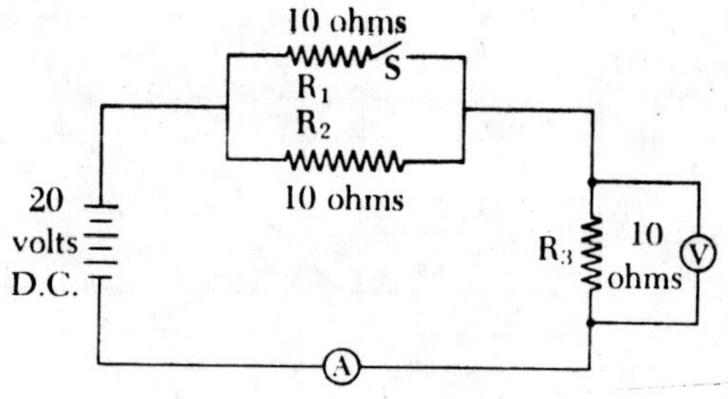

39. When switch S is open, the current through ammeter A is ____ ampere(s). 39.___
A. 0.5 B. 1.0 C. 2.0 D. 3.0

40. With switch S open, the voltage across resistor R_3 as read on voltmeter V is _____ volts. 40.___
A. 20 B. 15 C. 10 D. 5.0

41. If switch S is closed, the total resistance of the circuit is _____ ohms. 41.___
A. 10 B. 15 C. 20 D. 30

42. How will the final volume of a gas compare with its original volume if the pressure is doubled as the temperature is raised from 0°C to 273°C? 42.___
The final will be _____ the original.
A. $\frac{1}{4}$ B. $\frac{1}{2}$ C. the same D. 2 times

43. Of electromagnetic radiation, _____ has the SHORTEST wave-length. 43.___
A. gamma B. light C. infrared D. ultraviolet

44. When the rate of electron flow passing through a resistor of constant resistance is doubled, the voltage drop across the resistor is 44.___
A. doubled
B. quadrupled
C. halved
D. not affected

45. If an object is placed 10 centimeters from a convex (converging) lens of 5 centimeters focal length, the image formed will be 45.___
A. closer to the lens than 10 centimeters
B. larger than the object and virtual
C. farther from the lens than 10 centimeters
D. the same size as the object and inverted

Questions 46-47.

DIRECTIONS: Use the graph below in answering Questions 46 and 47.

The graph below shows resulting changes in temperature when heat is added at the constant rate of 20 calories per minute to 1.0 grams of a substance.

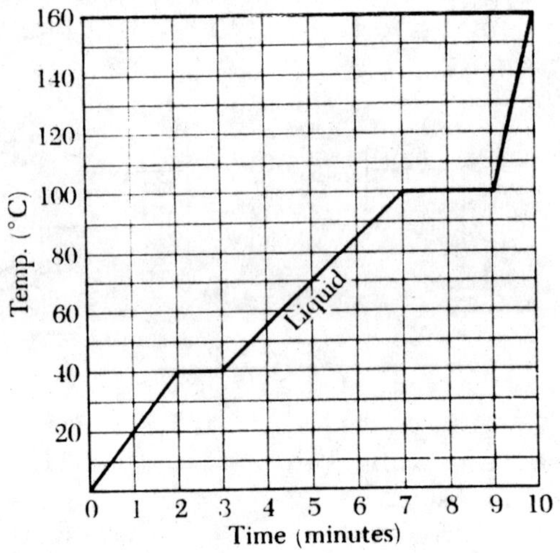

46. The heat of fusion of the substance is ____ calories per gram. 46.____

A. 2.0 B. 10 C. 20 D. 40

47. Compared with its heat of fusion, the substance's heat of vaporization is 47.____

A. three times as great B. twice as great
C. the same D. half as great

48. An energy level of a certain isolated atom is split into three components by the hyperfine interaction coupling of the electronic and nuclear angular momenta. The quantum number j, specifying the magnitude of the total electronic angular momentum for that level, has the value j = 3/2. The quantum number i, specifying the magnitude of the nuclear angular momentum, MUST have the value 48.____

A. i = ½ B. i = 1 C. i = 3/2 D. i = 2

49. 49.____

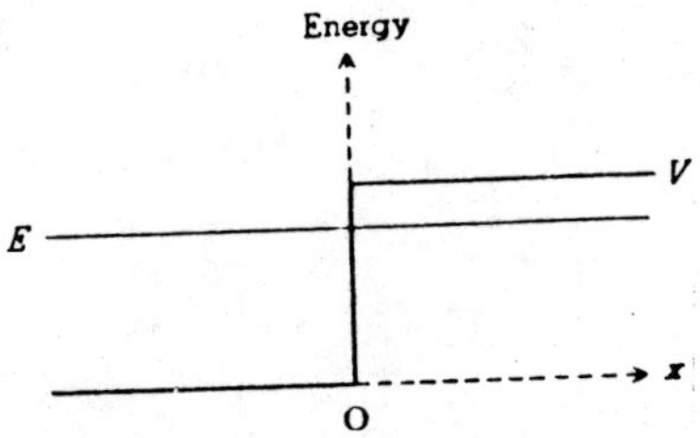

An electron with energy E and momentum kh is incident from the left on a potential step of height V > E at x = 0.
For x > 0, the space part of the electron's wave function has the form

A. e^{ikn}
B. $e^{-1k'x;k'<k}$
C. $e^{-\alpha x}$, where α is real and positive
D. sin kx

50. The nonconservation of parity in the decay $\pi^+ \rightarrow \mu^+ + v$ can be verified by measuring the 50.____

A. Q-value of the decay
B. longitudinal polarization of the μ^+
C. longitudinal polarization of the π^+
D. angular correlation between the μ^+ and the v

KEY (CORRECT ANSWERS)

1. E	11. A	21. A	31. C	41. B
2. C	12. B	22. B	32. C	42. C
3. B	13. E	23. B	33. D	43. A
4. C	14. D	24. B	34. A	44. A
5. C	15. B	25. A	35. D	45. D
6. A	16. D	26. D	36. D	46. C
7. C	17. A	27. C	37. B	47. B
8. D	18. C	28. B	38. B	48. B
9. A	19. B	29. D	39. B	49. C
10. E	20. C	30. C	40. C	50. B

TEST 2

DIRECTIONS: Each question or incomplete statement is followed by several suggested answers or completions. Select the one that BEST answers the question or completes the statement. *PRINT THE LETTER OF THE CORRECT ANSWER IN THE SPACE AT THE RIGHT.*

1. A four gram object floating on water has a buoyant force of — 1.____
   A. 4 gms.
   B. 3920 dynes
   C. 1/114 pounds
   D. cannot be determined

2. The height of a column of mercury compared to the height of a column of water measuring the atmospheric pressure would be — 2.____
   A. 13.6 times greater
   B. 1/13.6 times as great
   C. 4 times as great
   D. 1/4 times as great

3. In the equation p = hd, as h is doubled, p changes by a factor of — 3.____
   A. 2 B. 4 C. $\frac{1}{2}$ D. $\frac{1}{4}$

4. In the equation $E = \frac{1}{2}mv^2$, as v is doubled, E changes by a factor of — 4.____
   A. 2 B. 4 C. $\frac{1}{2}$ D. $\frac{1}{4}$

5. In the equation $I = cp/r^2$, as r is doubled, I changes by a factor of — 5.____
   A. 2 B. 4 C. 3 D. $\frac{1}{4}$

6. In the equation $F = mv^2/r$, as v is doubled, F changes by a factor of — 6.____
   A. 2 B. 4 C. $\frac{1}{2}$ D. $\frac{1}{4}$

Questions 7-12.

DIRECTIONS: Which of the following graphs BEST describe the situation described in Questions 7 to 12.

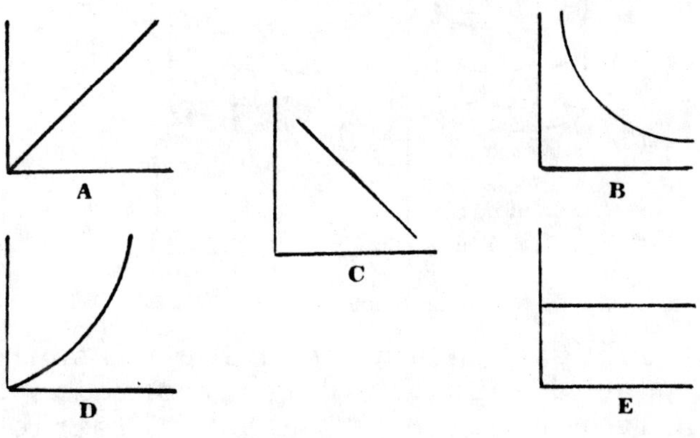

7. The velocity of the sweep hand of a watch versus time — 7.____
   A. A B. B C. C D. D E. E

8. The weight of objects in ounces versus corresponding weights of objects in pounds 8.____
A. A B. B C. C D. D E. E

9. The length of a candle versus the time it burns 9.____
A. A B. B C. C D. D E. E

10. Areas of circles versus diameters 10.____
A. A B. B C. C D. D E. E

11. A number b versus a number c so that bc gives a product of 12 11.____
A. A B. B C. C D. D E. E

12. The lengths of shadows versus the height of their respective opaque objects 12.____
A. A B. B C. C D. D E. E

13. An incoming light ray strikes a flat mirror and is reflected. 13.____
Which diagram BEST represents the path of this light ray?

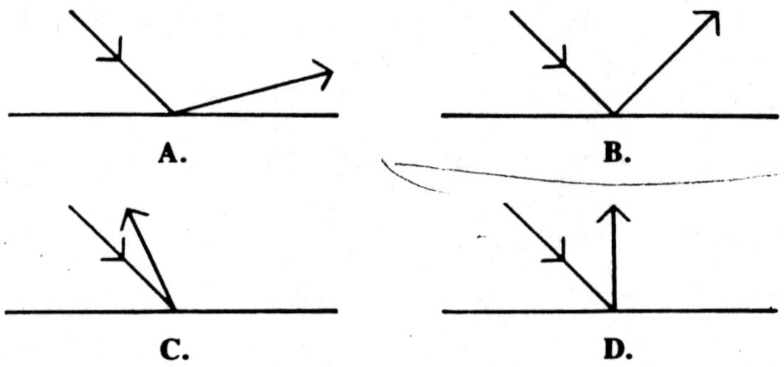

14. The efficiency of a pulley system is 14.____
A. rarely one
B. seldom five
C. always less than one
D. often greater than three

15. 15.____

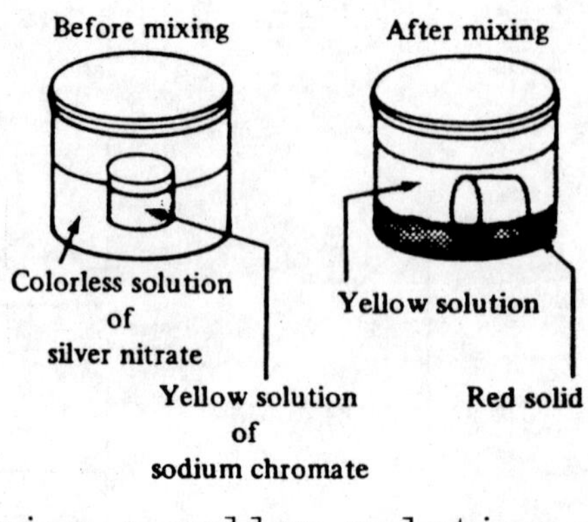

A beaker containing a yellow solution of sodium chromate is carefully placed into a jar that contains a colorless solution of silver nitrate. The jar is sealed, the mass of the system is determined, and the jar is slowly tipped over. When the two solutions come into contact, a thick red solid forms. The mass of the system is again measured. It is *probable* that the mass of the system after mixing would be

A. greater than the mass before mixing
B. less than the mass before mixing
C. the same as the mass before mixing
D. either greater or less than the mass before mixing, depending upon how heavy the red solid is

16. Which of these statements about sound is FALSE? 16.____

A. It *usually* involves a large amount of energy.
B. It ALWAYS originates with vibrations from a source.
C. A medium is needed to carry sound from source to receiver.
D. It travels faster than a speeding car in air, but much slower than light.

17. From the graph at the right, ____ grams of KNO_3 would dissolve in 100 grams of water at 50°C. 17.____

A. 20
B. 55
C. 68
D. 85

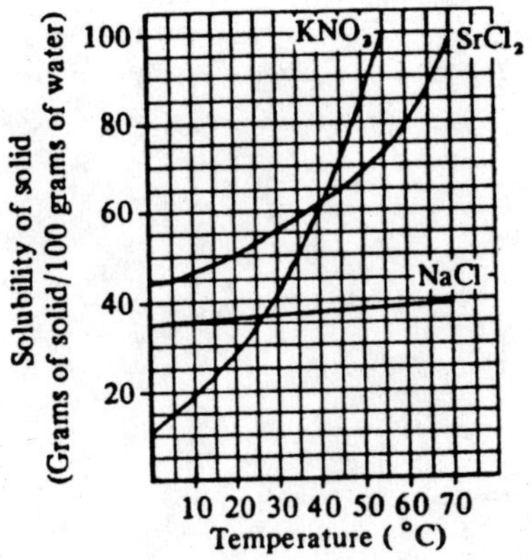

18. According to the Bohr model, the MAJOR force that holds an electron in an orbit about an atomic nucleus results from the 18.____

A. masses of the electron and the nucleus
B. charges of the electron and the nucleus
C. magnetic properties of the electron and the nucleus
D. kinetic energy of the electron in its orbit

19. In general, the speed of sound increases with an increase in the density of the medium that carries the sound. However, although cold air is denser than warm air, the speed of sound is *greater* in warm air than in cold air. Which is MOST important in explaining the speed of sound in air at different temperatures? 19.____

A. Molecules in warm air have higher average speeds than molecules in cold air have.
B. The wavelength of a sound wave through air becomes longer as the air is warmed.
C. Cold air is heavier than warm air.
D. The amplitude of a sound wave through air increases as the air temperature rises.

20. There is evidence that particles in addition to photons exhibit wave properties. However, there is *no* direct evidence that baseballs, for instance, have wave properties. An important reason for the absence of such evidence is that 20.____

A. the speed at which a baseball can be thrown is too slow for wave properties to be exhibited
B. the wavelengths for a moving baseball would be extremely short
C. only the fundamental particles can be expected to exhibit wave properties
D. only particles about to be converted to energy exhibit wave properties

21. Which of these shifts in energy level by an electron in a hydrogen atom results in the emission of the LARGEST amount of energy? 21.____

A. n = ∞ to n = 4
B. n = 4 to n = ∞
C. n = 1 to n = 3
D. n = 2 to n = 1

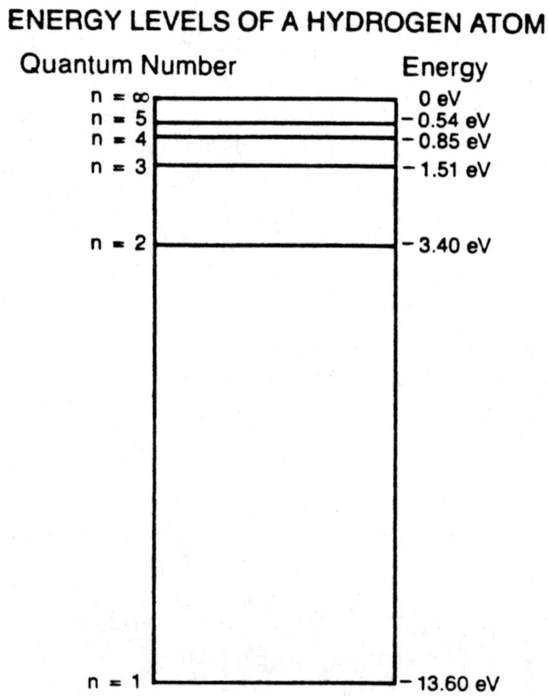

Questions 22-24.

DIRECTIONS: The following diagram shows a plot of the distance from the starting point *versus* time for an automobile after it starts from rest at a traffic light. Four points are marked on the plot. Use it to answer Questions 22 to 24.

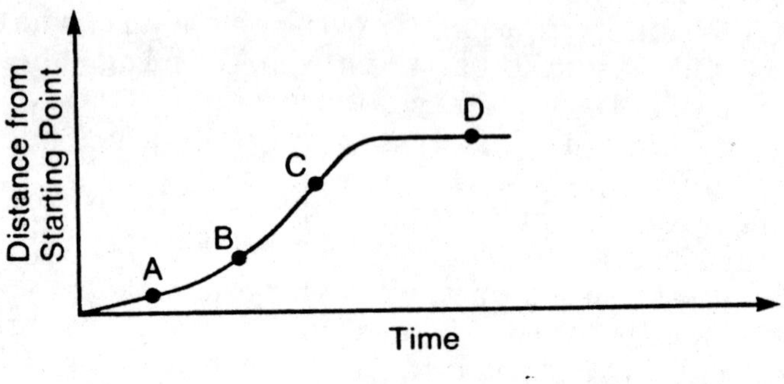

22. At what point is the car stopped? 22.____
A. A B. B C. C D. D

23. At what point is the speed of the car GREATEST? 23.____
A. A B. B C. C D. D

24. At what point does the car have a *positive* acceleration? 24.____
A. A B. B C. C D. D

25. A certain object slides on to a uniform level surface at an initial speed of 3 meters per second and comes to rest after traveling 1 meter. 25.____
If, instead, the initial speed of the object had been 6 meters per second, it would have come to rest after traveling ____ m.
A. 1 B. 2 C. 3 D. 4

26. Inverse proportionality is BEST illustrated by which of these relations? 26.____
A. Length of a spring and the magnitude of a weight that stretches the spring
B. Pressure and volume of a sample of gas kept at a constant temperature
C. Speed of an object and the time that a constant force acts on the object in the direction of motion of the object
D. Current and voltage in a closed circuit

27. According to the ideal gas law, an ideal gas at 0°K would have a *zero* value for all of the following properties EXCEPT 27.____
A. electrical resistance B. pressure
C. kinetic energy D. molecular motion

28. Despite the trajectory of a projectile, the force of gravity has the GREATEST effect when 28.____
A. the projectile has left the cannon
B. the projectile is at the top of its flight
C. just before the projectile strikes the ground
D. none of the above

29. A force with a magnitude of 2 newtons moves against an object making an angle of 30 degrees with the direction of motion. 29.____
The force causing the object to move is ____ nt(s).
A. 2 B. 1 C. $\sqrt{3}$ D. $\sqrt{2}$

30. An inclined plane making an angle of 30 degrees with the horizontal has a 10 newton object on it sliding down the incline with *no* acceleration. 30.____
The coefficient of friction is
A. 1 B. $\sqrt{3}$
C. $\frac{1}{2}$ D. none of the above

31. An object with a mass of 2 kg. and a velocity of 6 m/sec. has a momentum of _____ kg.m./sec. 31.____
   A. 12   B. 3   C. 8   D. 4

32. An object receiving a *constant* force of 6 newtons for 2 seconds has an impulse of _____ kg.m./sec. 32.____
   A. 12   B. 3   C. 8   D. 4

Questions 33-35.

DIRECTIONS: Questions 33 to 35 refer to the following:

Two carts with frictionless wheels and equal masses are capable of being loaded with various masses and are equipped for elastic collisions.

33. If an incident cart strikes a stationary target cart of equal mass, the motion following the collision will be 33.____
   A. both carts move in the same direction
   B. the incident cart stops, the target moves
   C. the incident cart bounces back, the target cart moves
   D. the incident cart bounces back, the target cart does not move

34. If a heavy incident cart strikes a lighter target cart, the following motion after the collision will result: 34.____
   A. both carts move in the same direction
   B. the incident cart stops, the target cart moves
   C. the incident cart bounces back, the target cart moves
   D. the incident cart bounces back, the target cart does not move

35. If a light incident cart strikes a heavier target cart, the following motion after the collision will result: 35.____
   A. both carts move in the same direction
   B. the incident cart stops, the target cart moves
   C. the incident cart bounces back, the target cart moves
   D. the incident cart bounces back, the target cart does not move

36. A 2 kg. mass receiving a 12 nt. force for one second will have a final velocity of _____ m/sec. 36.____
   A. 2   B. 12   C. 6   D. 24

37. A friction-free carload of sand has a velocity of 10 m/sec. and a leaky bottom. 37.____
As the sand leaks out, the velocity of the system will
   A. increase   B. decrease
   C. remain the same   D. be unpredictable

38. An object with a mass of 10 kg. falls a distance of 10 meters. 38.____
The kinetic energy of the particle immediately before impact is _____ joules.
   A. 10   B. 980   C. 100   D. 23.8

39. The velocity of the object immediately before impact was ____ m/sec. 39.____
A. 14 B. 14 C. 20 D. 10

40. The heat generated at the point of impact would be proportional to the 40.____
A. mass and the velocity
B. mass
C. mass and the square of the velocity
D. velocity

41. The change in potential energy as the object fell would be 41.____
A. unknown
B. equal to change in kinetic energy
C. 100 joules
D. $\sqrt{10}$ joules

42. A unit of measure of energy is the 42.____
A. horsepower B. watt C. calorie D. newton

43. A unit of measure of energy is the 43.____
A. kilogram B. B.T.U. C. newton D. watt

44. A unit of measure of force is the 44.____
A. $kg.m/sec^2$
B. slug
C. watt sec.
D. none of the above

45. A unit of measure of length is the 45.____
A. gram
B. cubic centimeter
C. liter
D. light year

46. A unit of measure of mass is the 46.____
A. newton
B. kilogram
C. newton meter
D. none of the above

Questions 47-48.

DIRECTIONS: Questions 47 and 48 are based on the following:

A particle with rest mass m and momentum mc/2 collides with a particle of the same rest mass that is initially at rest. After the collision, the original two particles have disappeared. Two other particles, each with rest mass m′, are observed to leave the region of the collision at equal angles of 30° with respect to the direction of the original moving particle, as shown below.

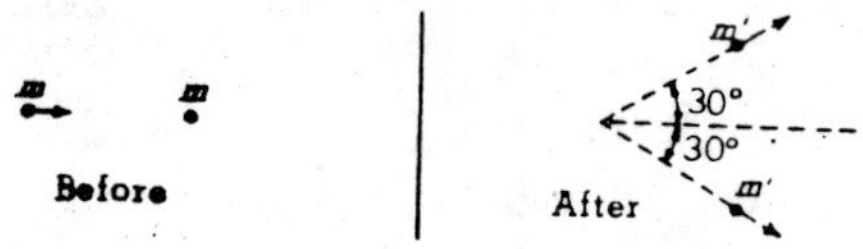

47. What is the speed of the original moving particle? 47.____
A. c/5 B. c/3 C. $c/\sqrt{7}$ D. $c/\sqrt{5}$

48. What is the momentum of each of the two particles produced by the collision? 48.____
   A. mc/5 B. $mc/2\sqrt{3}$ C. $mc/\sqrt{5}$ D. mc/2

Questions 49-50.

DIRECTIONS: Questions 49 and 50 are based on the following:

An ideal diatomic gas is initially at temperature T and volume V. The gas is taken through three reversible processes in the following cycle: adiabatic expansion to the volume 2V; constant volume process to the temperature T; isothermal compression to the original volume V.

49. For the complete cycle described above, which point is TRUE? 49.____
   A. Net thermal energy is transferred from the gas to the surroundings.
   B. Net work is done by the gas on the surroundings.
   C. The net work done by the gas on the surroundings is zero.
   D. The internal energy of the gas increases.

50. Which point about entropy changes in this cycle is TRUE? 50.____
   A. The entropy of the gas remains constant during each of the three processes.
   B. The entropy of the surroundings remains constant during each of the three processes.
   C. The combined entropy of the gas and surroundings remains constant during each of the three processes.
   D. For the complete cycle, the combined entropy of the gas and surroundings increases.

KEY (CORRECT ANSWERS)

1. B	11. B	21. D	31. A	41. B
2. B	12. A	22. D	32. A	42. C
3. A	13. B	23. C	33. B	43. B
4. B	14. C	24. B	34. A	44. A
5. D	15. C	25. B	35. C	45. D
6. B	16. D	26. B	36. C	46. B
7. E	17. D	27. A	37. C	47. D
8. A	18. B	28. D	38. B	48. B
9. C	19. A	29. C	39. B	49. A
10. D	20. A	30. D	40. C	50. C

EXAMINATION SECTION

TEST 1

DIRECTIONS: Each question or incomplete statement is followed by several suggested answers or completions. Select the one that BEST answers the question or completes the statement. *PRINT THE LETTER OF THE CORRECT ANSWER IN THE SPACE AT THE RIGHT.*

1. The metric system of measurement is used extensively in science because it 1.____
   A. is the only system which will work
   B. uses decimal notation
   C. is based on natural quantities
   D. is the only accurate system

2. The fact that one meter is NOT exactly one-tenmillionth the distance from the equator to the north pole 2.____
   A. should be corrected at some convenient time
   B. caused the gram to be an improper value
   C. is of no actual importance
   D. caused the liter to be an improper value

3. The meter is BEST defined as 3.____
   A. one-ten-millionth the distance from the equator to the north pole
   B. ten decimeters
   C. 39.97 inches
   D. the distance between two scratches on a standard bar

Questions 4-6.

DIRECTIONS: Questions 4 to 6 are based on the following:

A rectangular solid has a length of 6 cm., a width of 4 cm., and a height of 2 cm.

4. The volume of this solid is *nearly* _____ cubic inches. 4.____
   A. 48 B. 3 C. 19 D. 2.54

5. The height of the block is 5.____
   A. equal to an inch B. less than an inch
   C. greater than an inch D. close to .39 inches

6. The area of the SMALLEST side would be *nearly* ____ square inches. 6.____
   A. 8 B. 3 C. 1.2 D. 2.54

7. The operation involving a platform balance is probably MOST correctly called 7.____
   A. weighing B. massing C. forcing D. densing

8. The three fundamental quantities of physics would *probably* be considered as 8.____
   A. length, mass, and time
   B. length, area, and volume
   C. force, mass, and time
   D. force, mass, and length

9. The mass of 40 ml. of water would be about _____ grams. 9.____
   A. 20 B. 80 C. 60 D. 40

10. The number of millimeters in one meter is 10.____
   A. 1000 B. 1/1000 C. 100 D. 1/100

11. A system of measure on the moon compared to one on the earth could be the *same* 11.____
   A. as the earth's
   B. except time would have to be different
   C. except mass would have to be different
   D. except force would have to be different

12. A gallon is *approximately* the same volume as _____ liter(s). 12.____
   A. 1 B. 2 C. 3 D. 4

Questions 13-17.

DIRECTIONS: Questions 13 to 17 are based on the following:

A piece of wood is observed to be forcibly held on the bottom of a tank of water. The wood has a volume of 48 cubic centimeters and a mass of 35 grams.

13. The density of the wood is _____ cubic cm. 13.____
   A. 35 grams per B. 35 grams per 48
   C. 1.37 grams per D. none of the above

14. The buoyant force on the wood is 14.____
   A. 48 grams (47,000 dynes) B. 35 grams (34,400 dynes)
   C. 13 grams (12,700 dynes) D. zero (it is not floating)

15. The specific gravity of the wood is 15.____
   A. 35 B. 48 C. 1.37 D. .73

16. The wood *probably* has another force acting in a direction 16.____
   A. the same as the force of gravity
   B. opposite to gravity
   C. at right angles to gravity
   D. the same as the buoyant force

17. Since the object is at rest, this force 17.____
   A. cannot be determined
   B. has a value of 12,700 dynes
   C. has a value of 980 dynes
   D. has a value of 81,400 dynes

Questions 18-20.

DIRECTIONS: Questions 18 to 20 are based on the following:

A cube of wood is floating on top of water. The wood is known to have a mass of 18 grams and a volume of 27 cubic cm.

18. The buoyant force on this wood would be 18.____
    A. zero (it is not moving)
    B. 18 grams (17,600 dynes)
    C. 27 grams (26,400 dynes)
    D. cannot be determined with data given

19. This cube would be floating with this fractional part submerged: 19.____
    A. 2/3
    B. 1/3
    C. 1/2
    D. cannot be determined with data given

20. The volume of water displaced by the floating wood is ____ cu.cm. 20.____
    A. 18
    B. 27
    C. 9
    D. cannot be determined with data given

21. Using accepted notation, one million would be written 21.____
    A. 10^6 B. 10^{-6} C. 10^3 D. 10^{-3}

22. Using accepted notation, .001 would be written 22.____
    A. 10^{-2} B. 10^2 C. 10^{-3} D. 10^3

23. A balance will have the following relative positions of center of gravity and point of suspension: 23.____
    A. They will occupy the same point
    B. The center of gravity will be higher
    C. The center of gravity will be lower
    D. More than one of these will work

24. The ____ has the SMALLEST mass. 24.____
    A. neutron B. electron
    C. proton D. alpha particle

25. Which instrument listed below is the BEST one for directly measuring the mass of an object? 25.____
    A. Meter stick B. Graduated cylinder
    C. Beaker D. Balance

26. The liquid level in the graduated cylinder shown at the right is ____ milliliters. 26.____
    A. 22.0
    B. 22.3
    C. 22.6
    D. 23.2

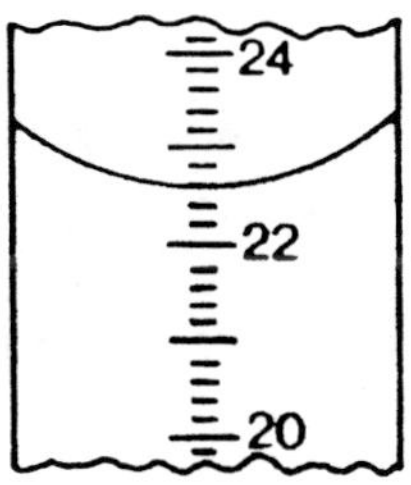

27. When a block of wood is lifted from the earth's surface to a point 100 kilometers up in the atmosphere, then the object's weight 27.____
   A. will decrease and its mass will increase
   B. will decrease and its mass will remain unchanged
   C. will remain unchanged and its mass will decrease
   D. and mass will remain unchanged

28. If the volume of a container holding a trapped gas is increased while the temperature of the gas is decreased, then the pressure the gas exerts will 28.____
   A. increase
   B. decrease
   C. remain unchanged
   D. decrease at first, then increase

29. 29.____

LINE	PARTICLE ARRANGE-MENT	PARTICLE MOTION	DISTANCE BETWEEN PARTICLES
A	particles held in definite pattern	particles vibrate back and forth	particles close together
B	no definite arrangement	particles move independently and freely	particles very far apart
C	no definite arrangement	particles vibrate back and forth	particles fairly close together
D	no definite arrangement	particles slip and slide over one another	particles fairly close together

Line ____ of the table above BEST describes the characteristics of a gas.
   A. A B. B C. C D. D

30. Why does tire pressure increase when air is pumped into an automobile tire? 30.____
   A. The tire expands, increasing the inside surface area.
   B. More air particles exert more force while the inside surface area remains almost constant.
   C. The force exerted per area of inside surface decreases when more particles are added.
   D. The force exerted by the particles inside the tire is reduced because the particles collide more with each other.

31. Which statement below about direct and alternating currents is TRUE? 31.____
Electrons flow
   A. in one direction in an alternating current; an alternating current is commonly produced by a generator

B. rapidly back and forth in an alternating current; an alternating current is commonly produced by a battery
C. rapidly back and forth in a direct current; a direct current is commonly produced by a battery
D. in one direction in a direct current; a direct current is commonly produced by a battery

32. An atom of one of the radioactive isotopes of sulfur emits a positron. Compared to the nucleus of this radioactive sulfur atom, the nucleus of the atom formed has _____ mass number and a _____ charge. 32.___
A. a larger; greater
B. the same; greater
C. the same; smaller
D. a smaller; smaller

33. To gather data on the speed of a toy car, one would find the MOST accurate to be a(n) 33.___
A. balance, a set of weights, and a meterstick
B. tape measure and a stop watch
C. yardstick and an oven timer
D. odometer (mileage indicator from the dashboard of a car) and an alarm clock

Questions 34-35.

DIRECTIONS: Questions 34 and 35 pertain to the firing of a rocket straight up from its launching pad.

34. All of these changes would make the rocket take off faster EXCEPT 34.___
A. firing more propellant out per second
B. firing the propellant out at greater speed
C. *decreasing* the mass of the rocket
D. *increasing* the mass of the rocket

35. The downward force on the rocket that is GREATEST before the rocket leaves its pad and decreases as the rocket rises is the 35.___
A. chemical reactive force
B. gravitational force
C. force of air friction
D. force exerted by the rocket engines

36. Physicists *usually* say that an atom of fluorine (atomic number 9, atomic weight 19) is composed of _____ protons, _____ electrons, and _____ neutrons. 36.___
A. 9; 9; 9
B. 9; 9; 10
C. 9; 10; 10
D. 10; 10; 9

37. In a constant-volume gas thermometer, a change in temperature is indicated by a change in the _____ of the gas. 37.___
A. pressure
B. density
C. number of molecules
D. specific heat

38. A force with a magnitude of 5 newtons gives to a certain body an acceleration of 4 meters per second squared. If the magnitude of the force is increased to 15 newtons, the acceleration of the body becomes _____ m/sec^2. 38.____

A. less than 4 B. 4
C. between 4 and 8 D. more than 8

39. In a particular power plant, fuel is burned to boil water. The steam is used to drive a turbine which in turn drives a generator. 39.____
Which energy transfer is NOT involved?

A. Mechanical to electrical
B. Electrical to chemical
C. Chemical to thermal
D. Thermal to mechanical

40. The two blocks of iron pictured at the right are identical in all ways EXCEPT that they are at different temperatures. 40.____

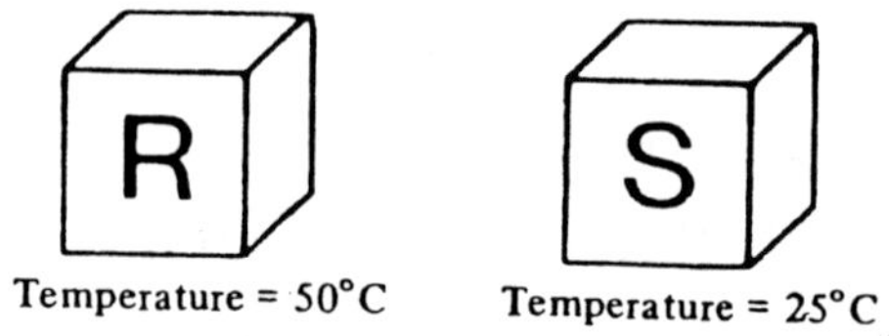

Which of the following statements is TRUE?

A. The particles in Block R are moving faster.
B. The particles in Block S are moving faster.
C. The particles in both blocks are moving at the same speed.
D. Temperature has no effect on particle motion.

41. A neutral atom of S becomes an S^{-2} ion by 41.____

A. gaining 2 electrons B. losing 2 electrons
C. gaining 2 protons D. losing 2 protons

42. In the formula Na_2SO_4, _____ total atoms are represented. 42.____

A. 3 B. 4 C. 6 D. 7

43. What will the CORRECT formula be when ions of Al^{+3} and Cl^{-1} are combined to make a neutral compound? 43.____

A. Al_3Cl B. $AlCl$ C. $AlCl_3$ D. $AlCl_2$

44. Elastic potential energy is present with a 44.____

A. coin sitting on top of a dresser
B. piece of elastic lying on the floor
C. stretched rubber band
D. paper airplane flying through the air

45. When placed into a graduated cylinder, a 12 gram rock causes the water level to go from 25 cm^3 to 30 cm^3. The density of the rock is _____ $grams/cm^3$. 45.____

A. 0.42 B. 0.48 C. 2.1 D. 2.4

46. 46.____

LINE	NUMBER OF PROTONS	NUMBER OF NEUTRONS	NUMBER OF ELECTRONS
A	9	9	9
B	9	10	9
C	9	19	10
D	19	9	19

A neutral atom of fluorine has an atomic number of 9 and a mass number of 19. Line _____ of the chart above *correctly* indicates the number of protons, neutrons, and electrons found in this atom.

A. A B. B C. C D. D

47. Which of these statements about acids is TRUE? 47.____

A. Acids corrode or react with active metals.
B. Acids cannot be neutralized by bases.
C. Acids change the color of litmus from red to blue.
D. When diluted, acids have a bitter taste.

48. There is NO kinetic energy in the case of a 48.____

A. golf ball rolling on the grass
B. stone skipping on top of the water
C. picture hanging on the wall
D. glass falling to the floor

Questions 49-50.

DIRECTIONS: Use the following diagram and data to answer Questions 49 and 50.

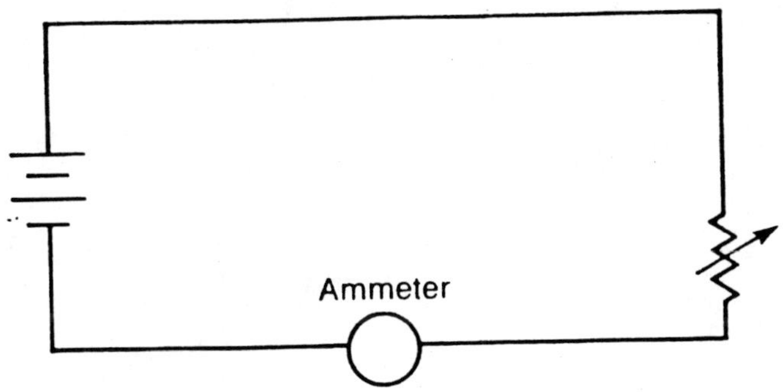

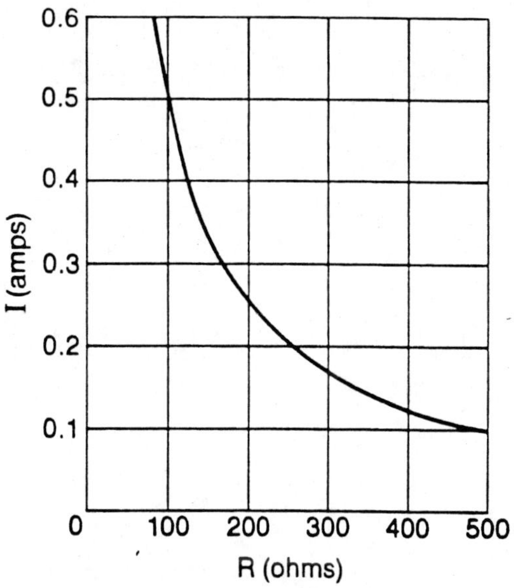

A variable resistor and an ammeter are connected in series to a battery that has electromotive force E. Data obtained by reading the current I from the ammeter for different values of resistance R are shown in the graph at the right above.

49. Which equation BEST represents this graph? 49.____
A. I = ER B. I = E+R C. I = E/R D. I = E-R

50. The constant value of E in this experiment is ____ V. 50.____
A. 10 B. 50 C. 100 D. 500

KEY (CORRECT ANSWERS)

1. B	11. A	21. A	31. D	41. D
2. C	12. D	22. C	32. C	42. D
3. D	13. B	23. C	33. B	43. C
4. B	14. A	24. B	34. D	44. C
5. B	15. D	25. D	35. B	45. D
6. C	16. A	26. C	36. B	46. B
7. A	17. B	27. B	37. A	47. A
8. A	18. B	28. B	38. D	48. C
9. D	19. A	29. B	39. B	49. C
10. A	20. A	30. B	40. A	50. B

TEST 2

DIRECTIONS: Each question or incomplete statement is followed by several suggested answers or completions. Select the one that BEST answers the question or completes the statement. *PRINT THE LETTER OF THE CORRECT ANSWER IN THE SPACE AT THE RIGHT.*

1. Work is BEST defined as 1.____
   A. force per unit distance
   B. force times distance
   C. product of force, distance, and the cosine of the angle between them
   D. product of force, distance, and the sine of the angle between them

2. Power is BEST defined as 2.____
   A. work per unit time
   B. product of work and time
   C. product of work, time, and the cosine of the angle between them
   D. product of work, time, and the sine of the angle between them

3. Efficiency of a machine is BEST defined as 3.____
   A. speed of doing work
   B. ratio of work out divided by work in
   C. ratio of work in divided by work out
   D. ratio of force out and force in

Questions 4-6.

DIRECTIONS: Questions 4 to 6 are based on the following:

A simple machine is made with a pry 8 ft. long with a support 1 ft. from the end. The support is six inches high.

4. This pry used as a first class lever would produce a MAXIMUM theoretical mechanical advantage of 4.____
   A. 7 B. 1/7 C. 8 D. 1/8

5. The pry used as a second class lever would produce a MAXIMUM theoretical mechanical advantage of 5.____
   A. 7 B. 1/7 C. 8 D. 1/8

6. The mechanical advantage of the first class lever would be increased by a factor of two if the 6.____
   A. support were twice as high
   B. support were one-half as high
   C. the long lever arm were increased by a length of two feet
   D. the short lever arm were decreased by a length of six inches

7. An oar in a rowboat is BEST described as a _____ class lever. 7.____
   A. first
   B. second
   C. third
   D. some combination of the above

8. A block and tackle is constructed using two movable pulleys and two fixed pulleys. 8.____
The MAXIMUM mechanical advantage with this system is
   A. 2   B. 4   C. 5   D. 10

9. An item below which is NOT an example of a wheel and axle is 9.____
   A. door knob   B. crank
   C. front wheel of car   D. back wheel of car

10. Given the equation: $y = kx^2$. As x is doubled, the value of y is increased by 10.____
   A. a factor of 2   B. 2
   C. a factor of 4   D. 4

11. Given the equation: $y = k/x$. As x is doubled, the value of y is 11.____
   A. increased by 2
   B. changed by a factor of one-half
   C. increased by 4
   D. decreased by a factor of $\frac{1}{4}$

12. Given the equation: $y = mx+b$, where b = zero. As x is doubled, the value of y is 12.____
   A. increased by 2
   B. increased by a factor of 2
   C. decreased by a factor of 2
   D. none of the above

13. Given the equation: $y = 3x$. As x is doubled, the value of y is increased by 13.____
   A. 2   B. 3
   C. a factor of 2   D. a factor of 3

Questions 14-19.

DIRECTIONS: Given below are some graphs which represent the relationships between variables obtained during various experiments. Use these graphs in answering Questions 14 to 19.

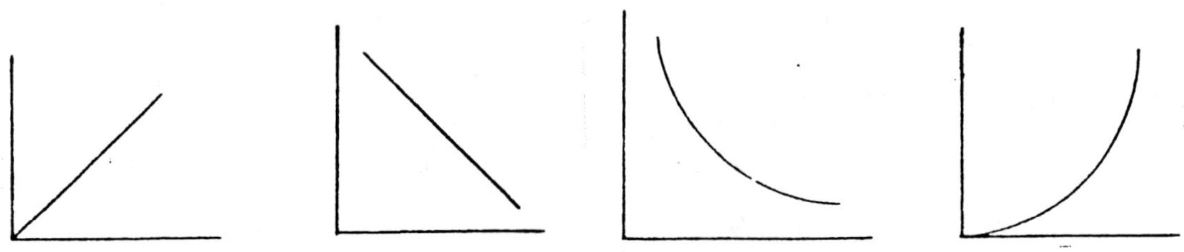

A   B   C   D

14. The graph obtained as a result of plotting time versus the distance traveled of a free falling object would be 14.____
 A. A B. B C. C D. D

15. The graph obtained as a result of plotting the stretch in a spring versus the distorting force would be 15.____
 A. A B. B C. C D. D

16. The graph obtained as a result of plotting pressure of a confined gas versus its volume would be 16.____
 A. A B. B C. C D. D

17. The graph obtained as a result of plotting time versus the velocity of a free falling object would be 17.____
 A. A B. B C. C D. D

18. The graph obtained as a result of plotting pressure versus the reciprocal of the volume of a confined gas would be 18.____
 A. A B. B C. C D. D

19. The graph obtained as a result of plotting n versus 1/n would be 19.____
 A. A B. B C. C D. D

20. A ten pound object is raised five feet in two minutes. The work done is 20.____
 A. 50 ft. lbs.
 B. 15 ft. lbs.
 C. 50 joules
 D. cannot be determined with data given

21. A ten pound object is raised five feet in two minutes. The power required is 21.____
 A. 50 watt hours A. 30 joules
 C. .972 horsepower D. none of the above

22. The efficiency of a machine is 80 percent. Twenty joules of energy is put into the machine. 22.____
The amount of work obtained would be _____ joules.
 A. 20
 B. 25
 C. 16
 D. cannot be determined with this data

23. An inclined plane 20 feet long makes an angle of 30° with the horizontal. A ten pound object is moved along its length. 23.____
The work done is _____ ft. lbs.
 A. 100 B. 200 C. 173 D. 30

24. Disregarding friction, the force necessary to move the object up the incline would be _____ lbs. 24.____
 A. 5 B. 10 C. 17.3 D. 8.66

25. If the coefficient of friction were 0.1, the angle at which an object would just slide the plane would be about _____ degrees. 25.____
A. 30 B. 60 C. 20 D. 6

Questions 26-29.

DIRECTIONS: Questions 26 to 29 pertain to waves in a pond of water. A top view of the wave crests in a section of the pond at one instant is shown below.

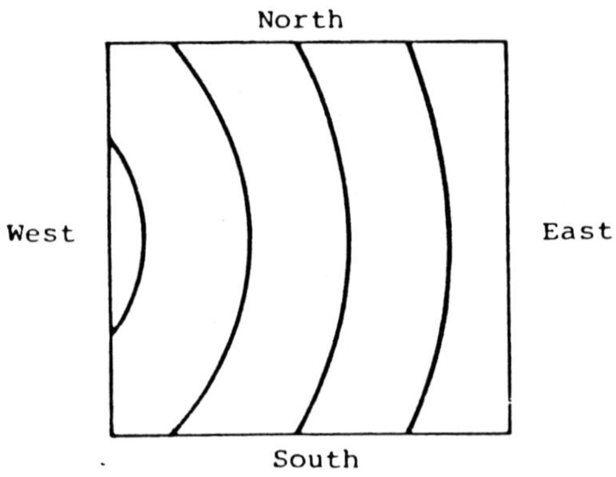

Figure 1

26. The wave at the left in the diagram above differs from the wave at the right since the 26.____
A. wave length is less
B. wave length is greater
C. amplitude is less
D. amplitude is greater

27. If a bigger rock had been dropped into the pond in the same spot and at the same time, the waves in the section shown would 27.____
A. be moving in the opposite direction
B. be moving faster
C. be moving slower
D. have higher crests and deeper troughs

28. If the wavelength of the waves is 1 meter and a new wave crest appears in the section each second, the speed of the waves is _____ m/sec. 28.____
A. 1/2
B. 1
C. 2
D. The speed cannot be determined from the information given.

29. 29.___

Figure 2

Suppose that a short time later, the wave crests on a section of the pond look for an instant like Figure 2 above.
Which fact could account for this observation?

A. A set of wave crests like that in Figure 1 was reflected from the shore of the pond.
B. Two rocks hit the pond at the same instant a whole number of wave lengths apart.
C. A set of wave crests like that in Figure 1 encountered two twigs sticking through the water exactly 1 wave length apart.
D. A set of wave crests like that in Figure 1 encountered a section of pond where the pond depth changes very rapidly.

30. A calorie can be defined as the amount of energy needed to raise the temperature of ____ Celsius degree(s). 30.___

A. 1 gram of water 1
B. 1 kilogram of water 100
C. 1 gram of any substance 1
D. 1 gram of oxygen 1

31. Given the following information: 31.___

Substance	Specific Heat of Substance $\left(\frac{\text{Calories}}{\text{gram Celsius degree}}\right)$
X	0.5
Y	1.0
Z	0.1

Which statement below would be a CORRECT conclusion if equal masses of each of these three substances absorbed the same amount of energy?

A. Substance X would heat up faster than Substance Z.
B. Substance Y would heat up the fastest.
C. Substance Y would heat up faster than Substance Z.
D. Substance Z would heat up faster than Substance Y.

32. ____ calorie(s) would be required to heat 30 grams of water from 10°C to 50°C. 32.____

A. 1 B. 300 C. 1200 D. 1500

33. If an atom were composed of 1 proton, 1 neutron, and 1 electron, it would *probably* 33.____

A. be neutral and have a mass of 3 amu.
B. have a charge of +1 and have a mass of 0 amu.
C. be neutral and have a mass of 2 amu.
D. be neutral and have a mass of 0 amu.

34. If you were trying to separate the dissolved salt from the water in a sample of seawater, the BEST procedure would be to 34.____

A. filter the seawater through a piece of nylon stocking
B. filter the seawater through a piece of very fine filter paper
C. distill the solution by catching the recondensed water in another container while leaving the solid salt behind
D. distill the solution first, pour the recondensed water back into the container with the solid salt, then filter it through fine filter paper

35. In each choice below, a method of heat transfer is given along with an example of that type of heat transfer. In which case below is the example NOT a CORRECT match for the given method of heat transfer? 35.____

A. Convection: warming hands by holding them in front of a fire
B. Conduction: heating a pan on a stove
C. Radiation: the earth being heated by the sun
D. Conduction: a metal poker gets hot when one end is put into a fire

36. 36.____

LINE	Was the solid crushed into smaller pieces?	Were the solid and water stirred together?	What was the temperature of the water?
A	Yes	No	30°C
B	Yes	Yes	30°C
C	Yes	Yes	50°C
D	No	Yes	50°C

Line ____ of the chart above illustrates the conditions in which a given solid would *probably* dissolve MOST rapidly in water.

A. A B. B C. C D. D

37. Solid materials which are good conductors of electricity have many loosely bound 37.___
A. neutrons B. protons C. electrons D. nuclei

38. Atoms that have the same atomic number but different mass numbers are 38.___
A. isotopes B. isobars C. isotherms D. isomers

Question 39.

DIRECTIONS: Use the diagram below in answering Question 39.

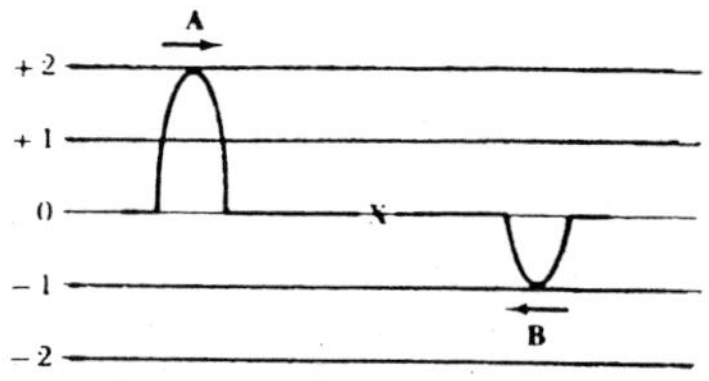

39. Two pulses, A and B, approach each other along a rope as shown in the diagram above. 39.___
Which of the following diagrams indicates the resulting wave form when the peak amplitudes of both pulses are at point X?

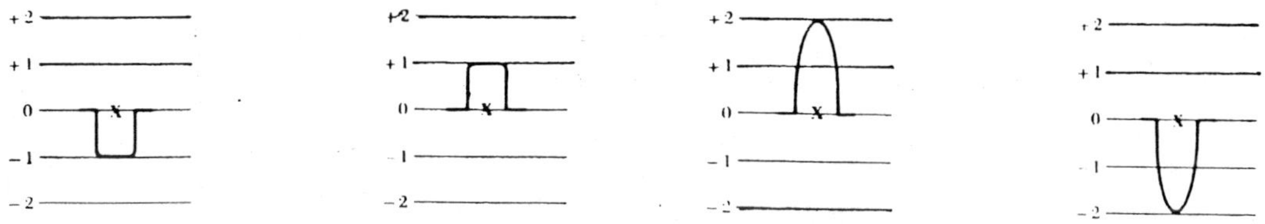

A. B. C. D.

40. As the acceleration of a moving object (assume constant mass) is doubled, the force required to impart this acceleration 40.___
A. is halved B. is doubled
C. is quadrupled D. remains the same

41. Element Q decays to form element X according to the following reaction: ${}_{84}Q^{210} \rightarrow X + {}_{2}He^{4}$ 41.___

The mass number of element X is
A. 206 B. 208 C. 212 D. 214

42. A light ray is emitted from the flame of a candle at C and strikes a plane mirror at A. To an observer at O, the ray will seem to come from point 42.___
A. A
B. B
C. C
D. D

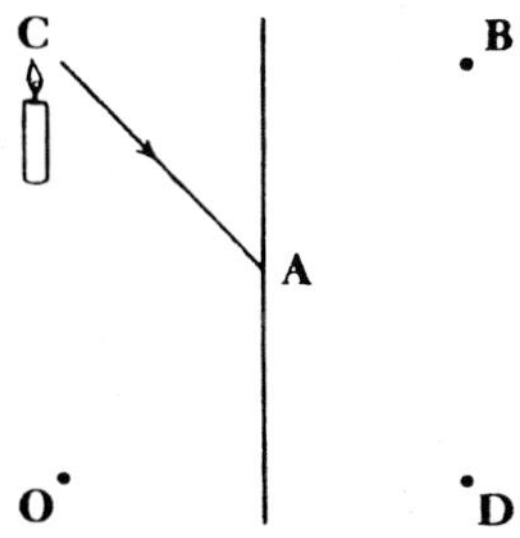

43. The polarity of the ends of an electromagnet depends upon the 43.____
A. amount of the current in the coil
B. number of turns of wire in the coil
C. material of the core
D. direction of the current in the coil

44. A man lifts a 5 pound weight a vertical height of 5 feet in 10 seconds. A second man lifts an equal weight to the same height in 20 seconds. 44.____
The work done by the first man compared with the work done by the second man is
A. half as much B. the same
C. twice as much D. four times as much

Questions 45-47.

DIRECTIONS: The table below presents the results of an experiment in which different masses of water were heated for different lengths of time producing various changes in the temperature. Use the table and formula to answer Questions 45 to 47.

TRIAL	MASS OF WATER	TIME HEATED	TEMPERATURE CHANGE
1	100 grams	1 min.	10°C
2	100 grams	2 min.	19°C
3	200 grams	1 min.	5°C
4	200 grams	2 min.	11°C
5	300 grams	5 min.	16°C
6	300 grams	10 min.	34°C

There is a formula, $H = m \times s.h. \times \Delta t$, where
H = heat added to some water,
m = mass of the water,
$s.h.$ = specific heat of the water, a constant, and
Δt = the temperature change of the water

45. Which set of apparatus would *probably* be MOST useful in performing this experiment? 45.____
A. Balance, beaker, burner, thermometer
B. Compass, ruler, balance, thermocouple
C. Cylinder, flask, galvanometer, thermometer
D. Battery, mirror, pendulum, wire

46. To see how well the numbers in the table fit the formula, it would be MOST important to make which of these assumptions? 46.____
A. The same amount of heat was supplied during each minute of heating.
B. All six trials were started at the same time and went on together.
C. Six different thermometers were used, one for each trial.
D. The experiment was carried out at sea level.

47. If 200 grams of water had been heated for 5 minutes, what would have been the *approximate* temperature change of the water? 47.___

A. 4 C
B. 15 C
C. 25 C
D. No estimate can be made from the information given

Questions 48-50.

DIRECTIONS: Questions 48 to 50 relate to a particle of mass M that is moving in an attractive central force field. The potential function representing the attractive central force field can be written as $V(r) = -K/r$. At a certain time, the particle has angular momentum L and total energy E.

48. At some later time, which of these statements will be TRUE of the angular momentum L and total energy E of the particle? 48.___

A. L will have changed but E will not.
B. E will have changed but L will not.
C. Neither L nor E will have changed.
D. Both L and E will have changed.

49. For a given nonzero angular momentum, there is a MINIMUM energy for which it is possible to find a solution to the equations of motion. 49.___
At this MINIMUM energy, the particle is moving in a ____ orbit.

A. circular
B. noncircular elliptical
C. parabolic
D. hyperbolic

Question 50.

DIRECTIONS: Question 50 concerns a uniformly charged wire that has the form of a circular loop with radius b. Consider two points on the axis of the loop. P_1 is at a distance b from the loop's center, and P_2 is at a distance 2b from the loop's center. The potential V is zero very far from the loop. At P_1 and P_2, the potentials are V_1 and V_2, respectively.

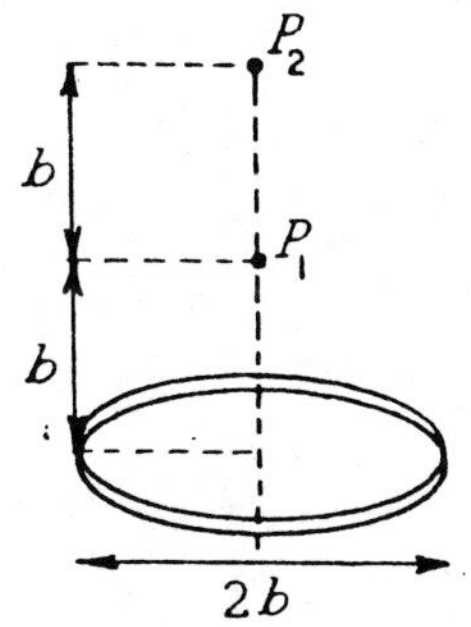

50. What is V_2 in terms of V_1? 50.___

A. $\frac{V_1}{3}$
B. $\frac{2V_1}{5}$
C. $\frac{V_1}{2}$
D. $\sqrt{\frac{2}{5}}\, V_1$

KEY (CORRECT ANSWERS)

1. C	11. B	21. D	31. D	41. A
2. A	12. B	22. C	32. C	42. B
3. B	13. C	23. A	33. C	43. D
4. A	14. D	24. A	34. C	44. B
5. C	15. A	25. D	35. A	45. A
6. D	16. C	26. D	36. C	46. A
7. B	17. A	27. D	37. C	47. C
8. C	18. A	28. B	38. A	48. C
9. C	19. C	29. B	39. B	49. A
10. C	20. A	30. A	40. B	50. D

TEST 3

DIRECTIONS: Each question or incomplete statement is followed by several suggested answers or completions. Select the one that BEST answers the question or completes the statement. *PRINT THE LETTER OF THE CORRECT ANSWER IN THE SPACE AT THE RIGHT.*

1. Three undefined fundamental units of measure used in physics are 1.____
   A. mass, weight, time
   B. mass, length, time
   C. length, time, area
   D. volume, area, length

2. The metric system of measure is *especially* desirable because 2.____
   A. of its natural units
   B. it is more accurate
   C. of its decimal notation
   D. of its universal adaptation

3. An object that measures 15 cm. long has an *approximate* length of _____ inches. 3.____
   A. 15
   B. 6
   C. 38
   D. .0392

4. An object of Area A is moved through space a distance d. The volume inscribed is 4.____
   A. Ad
   B. d^3
   C. A/d
   D. $Ad^2\pi$

5. If our units of measure were introduced on Venus, the one which would have to be changed would be 5.____
   A. mass
   B. force
   C. length
   D. none of the above

6. If a unit of volume were a lag, it would be convenient to have a unit of mass the marg so that the density of water would be _____ marg/lag. 6.____
   A. 1
   B. 2.54
   C. 12
   D. none of the above

7. The MOST accurate measurement a physicist can now make is *probably* about one part in 7.____
   A. a hundred
   B. a million
   C. ten million
   D. a billion

8. The length of a rectangular box is measured to be 3.2 inches, the height 4.0 inches, and the width 3.25 inches. The volume would be CORRECTLY expressed as 8.____
   A. 41.600 $inches^3$
   B. 42 $inches^3$
   C. 41.6 $inches_3$
   D. none of the above

9. The density of an object is found to be 7.64 gm/cm^3. The accepted value is 7.54 gm/cm^3. The percent of error is 9.____
   A. 1.33
   B. 1
   C. .75
   D. 7.5

10. The density of a cylinder is found with the following data: mass = 2.6538 gms., length = 6.7 cm., diameter = 0.13 cm. 10.____
To take better data, an improvement should be made in measuring
A. length B. mass C. diameter D. pi

11. An object that floats in water tells us that its specific gravity is 11.____
A. greater than one
B. less than one
C. is equal to one
D. more than one of the above

12. If an object has a mass of 26 gms. and a volume of 30 cm^3, its density is _____ gm/cm^3. 12.____
A. 4 B. 56 C. .868 D. 1.15

13. A column of water four feet high with a base of one-half ft^2 would exert a pressure of _____ $lbs./in^2$. 13.____
A. 8 B. 6.93 C. 4 D. 1.7

14. Which of the measurements below is NOT a value of standard pressure? 14.____
A. 13.6 lbs/in^2 B. 76 cm.Hg
C. 1 atmosphere D. 29.92 inches of Hg

15. Atmospheric pressure is *usually* measured with a 15.____
A. barometer B. thermometer
C. hygrometer D. airometer

16. The phenomenon of Pascal's Law is illustrated in a(n) 16.____
A. floating wood B. wind mill
C. hydraulic jack D. wheelbarrow

17. A floating iceberg has the following fractional part of its volume below the surface of the water: 17.____
A. one tenth B. nine tenths
C. one third D. two thirds

Questions 18-21.

DIRECTIONS: Questions 18 to 21 refer to the following:

A force of 2 newtons is acting on a 4 kilogram object. Friction is to be disregarded, and the force is parallel to the direction of motion.

18. If the distance moved is 3 meters, then the work done is 18.____
A. 6 joules B. 12 kg. meters
C. 3/2 joules D. 4/3 kg. meters

19. The acceleration of the object will be _____ m/sec^2. 19.____
A. 2 B. $\frac{1}{2}$
C. 9.8 D. will not be a constant

20. The power developed by this system would be 20.____
A. 2 watts
B. 2 joules/sec
C. 1/15 horsepower
D. cannot be determined with data given

21. The kinetic energy of the object after it has moved six meters is 21.____
A. 12 joules
B. 24 kg. meters
C. 18 orgs
D. cannot be determined

22. A free-falling object that had dropped for three seconds would have a velocity of 22.____
A. 9.8 m/sec.
B. 29.4 m/sec.
C. 32 ft/sec.
D. 25.4 m/sec.

23. A free-falling object that had dropped for three seconds would have traveled a distance of _____ meters. 23.____
A. 44.2 B. 9.8 C. 29.4 D. 32

Questions 24-26.

DIRECTIONS: Questions 24 to 26 refer to the following:

A pendulum with a mass of 2 kilograms is suspended from a massless rod 1 meter long.

24. The pendulum has a MAXIMUM velocity when the 24.____
A. kinetic energy is a minimum
B. potential energy is a maximum
C. acceleration is a maximum
D. potential energy is a minimum

25. The period of the pendulum would be increased if the 25.____
A. rod were lengthened
B. mass were increased
C. mass were decreased
D. amplitude were increased

26. The motion of the pendulum swinging with small amplitude is BEST described by the term _____ motion. 26.____
A. simple harmonic
B. repetitive
C. sine wave
D. oscillatory

27. The quantity listed below which is NOT a vector quantity is 27.____
A. force B. torque C. displacement D. time

28. A vector of 6 units north is added to a vector of 8 units south. 28.____
The resultant vector is one of
A. 2 units south
B. 14 units
C. 10 units west
D. none of the above

29. The density of a piece of coal weighing 6 grams and having a volume of 4 cubic centimeters is 29.____
A. 2/3 gm.
B. 2/3 cu. cm.
C. 1.5 gm.
D. 1.5 gm. per cu. cm.

30. Which of these terms is unrelated to the others? 30.____
A. Moment B. Fulcrum C. Lever D. Momentum

31. The ideal mechanical advantage of the pulley system shown at the right is 31.____
A. 1/4
B. 1/3
C. 2
D. 3

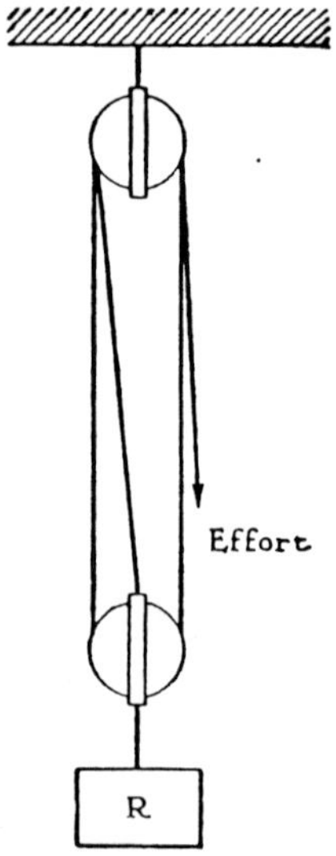

32. The hydraulic press is an application of a principle proposed by 32.____
A. Archimedes B. Boyle C. Charles D. Pascal

33. The pressure was determined at a depth of one foot in a liquid with a specific gravity of 1.2. 33.____
The same pressure in a liquid with a specific gravity of 0.6 would be found at a depth of ____ ft.
A. 0.5 B. 0.6 C. 1.0 D. 2.0

34. When one force of 12 pounds and another force of 5 pounds act at right angles to each other on a point, the resultant force is ____ lb. 34.____
A. 2.4 B. 7.0 C. 13.0 D. 17.0

35. A large steel drum containing air at normal atmospheric pressure is found to float with 40 percent of its volume under water. Compressed air is forced into the drum until the pressure is doubled. 35.____
The drum will
A. float higher in the water
B. sink to the bottom
C. still float with 40 percent of its volume under water
D. float lower in the water

36. To compute the power necessary to move a given body a given distance by a given force, one MUST also know the ____ the body. 36.____
A. time required to move B. mass of
C. velocity of D. acceleration given

37. An automobile starting from rest is given a constant acceleration of 4 feet per second for 10 seconds. During these 10 seconds, its average speed is ____ ft. per sec. 37.____
A. 0.4 B. 2.5 C. 4.0 D. 20.0

38. A machine may be used to multiply 38.____
A. force and speed
B. work
C. energy
D. force only or speed only

39. The outer pail of an ice cream freezer is *usually* made of wood because wood 39.____
A. has a low specific heat
B. is a poor conductor of heat
C. is a poor radiator of heat
D. is waterproof

40. When the bulb of a glass thermometer is plunged into hot water, the mercury in the stem first falls before rising because the 40.____
A. glass has a greater coefficient of cubical expansion than the mercury
B. mercury first contracts
C. glass first absorbs heat from the mercury
D. glass starts to expand before the mercury

41. Silvered surfaces are used in a vacuum bottle because silver 41.____
A. reflects heat waves
B. is a good radiator of heat
C. has a higher coefficient of expansion than glass has
D. is a poor conductor of heat

42. 42.____

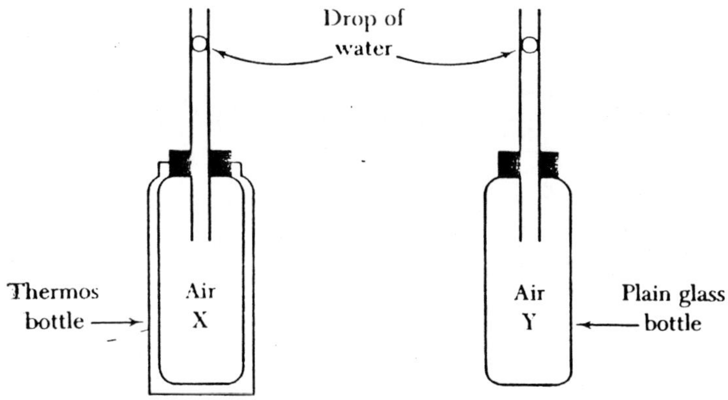

Shown above are two bottles containing equal volumes of air. When the bottles are taken from room temperature and placed in a refrigerator, the drop of water will move
A. down faster in X than in Y
B. up faster in Y than in X
C. up faster in X than in Y
D. down faster in Y than in X

43. The electron-volt is a unit of 43.____
A. power B. force
C. field intensity D. energy

44. If the gas in a rigid container is cooled but remains in a gaseous state, there will be a change in the ____ of the gas molecules. 44.____
A. volume B. average kinetic energy
C. average weight D. mass

45. If the distance between two spheres of constant mass is doubled, the gravitational force between them changes from F to 45.____
A. 4F B. 2F C. $\frac{1}{4}$F D. $\frac{1}{2}$F

46. Which of the following is a *scalar* quantity? 46.____
A. Force B. Acceleration C. Mass D. Momentum

Questions 47-48.

DIRECTIONS: Use the graphs below in answering Questions 47 and 48.

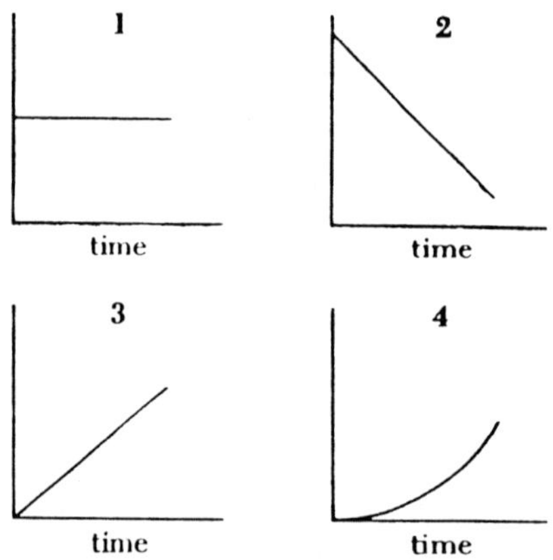

47. Which graph represents the relationship between time and speed of a car which is undergoing uniform *positive* acceleration? 47.____
A. 1 B. 2 C. 3 D. 4

48. Which graph represents the relationship between time and the distance covered by a freely falling stone? 48.____
A. 1 B. 2 C. 3 D. 4

49. If a light ray A strikes the air-water surface at an angle greater than the critical angle (as shown in the diagram at the right), light ray A will continue as ray 49.____
A. 4
B. 3
C. 2
D. 1

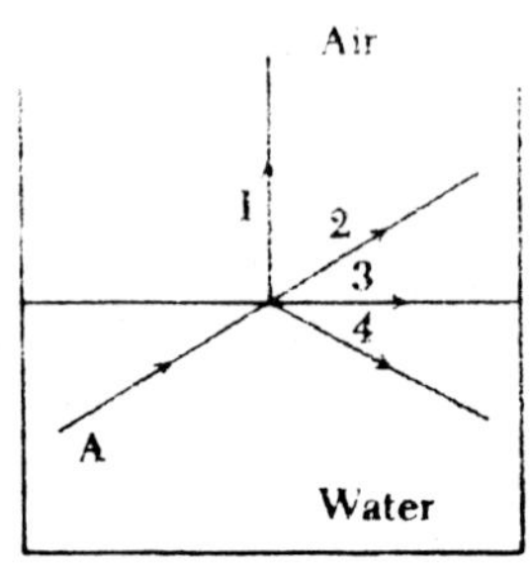

50. An isotope of polonium has a half life of 140 days. If a physicist had a sample of one gram of polonium, he could expect to have _____ gram remaining after 560 days. 50.___

A. 1/4 B. 1/8 C. 1/16 D. 1/32

KEY (CORRECT ANSWERS)

1. B	11. B	21. A	31. D	41. A
2. C	12. C	22. B	32. D	42. D
3. B	13. D	23. A	33. D	43. D
4. A	14. A	24. D	34. C	44. B
5. D	15. A	25. A	35. D	45. C
6. A	16. C	26. A	36. A	46. C
7. C	17. B	27. D	37. D	47. C
8. B	18. A	28. A	38. D	48. D
9. A	19. B	29. D	39. B	49. A
10. C	20. D	30. D	40. D	50. C

75 PHYSICAL FORMULAS

CONTENTS

CONTENTS (cont'd)

75 PHYSICAL FORMULAS

1. WEIGHT DENSITY

$$D = \frac{w}{V}$$

D is density; w is weight; V is volume

2. LIQUID PRESSURE

$p = hD$

p is pressure; h is depth; D is density

3. TOTAL FORCE

(1) Horizontal Surfaces

$F = AhD$

F is total force; A is area; h is depth; D is density

(2) Vertical Surfaces

$$F = \frac{AhD}{2}$$

F is total force; A is area; h is depth; D is density

4. MECHANICAL ADVANTAGE OF ANY MACHINE

$$\text{Mechanical Advantage} = \frac{w}{F}$$

w is weight supported; F is force applied

5. MECHANICAL ADVANTAGE OF HYDRAULIC PRESS

$$\text{Mechanical Advantage} = \frac{w}{F} = \frac{A}{a} = \frac{D}{d}$$

w is weight supported; F is force applied; A is area of large piston; a is area of small piston; D is diameter of large piston; d is diameter of small piston

6. SPECIFIC GRAVITY OF A SOLID OR LIQUID

$$\text{Sp. gr.} = \frac{\text{density of substance}}{\text{density of water}}$$

7. SPECIFIC GRAVITY OF A SOLID

(1) Denser than Water

$$\text{Sp. gr.} = \frac{\text{weight in air}}{\text{buoyant force of water}}$$

(2) Less Dense than Water

$$\text{Sp. gr.} = \frac{w}{w' - w''}$$

w is weight of solid in air; w' is combined weight of solid in air and sinker in water; w'' is combined weight of both solid and sinker in water

8. SPECIFIC GRAVITY OF A LIQUID

(1) Bottle Method

$$\text{Sp. gr.} = \frac{\text{weight of liquid}}{\text{weight of water}}$$

(2) Loss-of-Weight Method

$$\text{Sp. gr.} = \frac{\text{buoyant force of liquid}}{\text{buoyant force of water}}$$

(3) Hydrometer Method

$$\text{Sp. gr.} = \frac{\text{depth rod sinks in water}}{\text{depth rod sinks in liquid}}$$

9. BOYLE'S LAW

$$pV = p'V'$$

p is original pressure; V is original volume; p' is new pressure; V' is new volume

10. HOOKE'S LAW

$$\text{Elastic Modulus} = \frac{\text{stress}}{\text{strain}}$$

11. FACTOR OF SAFETY

$$\text{Factor of Safety} = \frac{\text{maximum load}}{\text{rated load}}$$

12. COMPOSITION OF FORCES

The resultant of two forces acting at an angle upon a given point is equal to the diagonal of a parallelogram of which the two force vectors are sides. The equilibrant equals the magnitude of the resultant, but acts in the opposite direction.

13. RESOLUTION OF FORCE OF GRAVITY

Object Resting on Inclined Plane
$W : W_p = l : h$

W is weight of object; W_p is force tending to pull object down plane; l is length of plane; h is height of plane

14. COEFFICIENT OF FRICTION

$$u = \frac{f}{N}$$

u is coefficient of friction; f is force of friction; N is force normal to surface

15. SPEED

$$\text{Average Speed} = \frac{\text{distance traveled}}{\text{elapsed time}}$$

16. ACCELERATED MOTION

$v = at$, or $v = gt$
v is final velocity; a is acceleration, or g is acceleration due to gravity; t is time

17. ACCELERATED MOTION

$s = \frac{1}{2}at^2$, or $s = \frac{1}{2}gt^2$
s is total distance; a is acceleration, or g is acceleration due to gravity; t is time

18. ACCELERATED MOTION

$v = \sqrt{2as}$, or $v = \sqrt{2gs}$
v is final velocity; a is acceleration, or g is acceleration due to gravity; s is total distance

19. ACCELERATED MOTION

$s = \frac{1}{2}a(2t-1)$, or $s = \frac{1}{2}g(2t-1)$
s is distance traversed in a given second; a is acceleration, or g is acceleration due to gravity; t is the number of the given second

20. NEWTON'S SECOND LAW OF MOTION

$F = ma$
F is force; m is mass; a is acceleration

21. FORCE AND ACCELERATION ON BODIES OF KNOWN WEIGHT

$F : w = a : g$
F is force; w is weight; a is acceleration; g is acceleration due to gravity

22. IMPULSE AND MOMENTUM

$Ft = mv$
F is force; t is time; the product Ft is impulse; m is mass; v is velocity; the product mv is momentum

23. CENTRIFUGAL FORCE

$$\text{Centrifugal Force} = \frac{mv^2}{r}$$

m is mass; v is velocity; r is radius of path

24. PENDULUM

$t : t' = \sqrt{l} : \sqrt{l'}$

t is period of first pendulum; t' is period of second pendulum; l is length of first pendulum; l' is length of second pendulum

25. PENDULUM

$$t = 2\pi\sqrt{\frac{l}{g}}$$

t is period; l is length; g is acceleration due to gravity

26. WORK

$W = Fs$

W is work; F is force; s is distance

27. POWER

$$P = \frac{W}{t}$$

P is power; W is work; t is time

28. HORSEPOWER

$$hp = \frac{Fs}{550t}$$

hp is horsepower; F is force in pounds; s is distance in feet; t is time in seconds

29. POTENTIAL ENERGY

$P.E. = mgh$

P.E. is potential energy; m is mass; g is acceleration due to gravity; h is vertical distance

30. KINETIC ENERGY

$K.E. = \frac{1}{2}mv^2$

K.E. is kinetic energy; m is mass; v is velocity

31. VARIATION OF MASS WITH VELOCITY

$$m = \frac{m_0}{\sqrt{1 - \frac{v^2}{c^2}}}$$

m is mass at velocity, v; m_0 is mass at zero velocity; c is speed of light

32. MACHINES

$F \times s_f = w \times s_w$

F is acting force; s_f is distance acting force moves; w is resisting weight; s_w is distance resisting weight moves

33. MACHINES

$$\text{Efficiency} = \frac{\text{useful work}}{\text{total work}}$$

34. LEVER

$$\text{Mechanical Advantage} = \frac{EF}{RF}$$

EF is length of effort arm; RF is length of resistance arm

35. PULLEY

$nF = w$
n is number of strands supporting movable block; F is acting force; w is resisting weight

36. WHEEL AND AXLE

$$\text{Mechanical Advantage} = \frac{C}{c} = \frac{D}{d} = \frac{R}{r}$$

C, D, and R are circumference, diameter, and radius, respectively, of wheel; c, d, and r are circumference, diameter, and radius, respectively, of axle

37. INCLINED PLANE

(1) When Force Acts Parallel to Plane

$$\text{Mechanical Advantage} = \frac{l}{h}$$

l is length of plane; h is height of plane

(2) When Force Acts Parallel to Base of Plane

$$\text{Mechanical Advantage} = \frac{b}{h}$$

b is base of plane; h is height of plane

38. SCREW

$$\text{Mechanical Advantage} = \frac{2\pi r}{d}$$

r is length of arm on which force acts; d is pitch of screw

39. COMPOUND MACHINE

Total Mechanical Advantage = mechanical advantage (machine 1) x mechanical advantage (machine 2) x mechanical advantage (machine 3), etc.

40. WORM WHEEL

Mechanical Advantage = $\frac{nl}{r}$

n is number of teeth in gear wheel; l is radius of wheel on which force acts; r is radius of axle on which weight acts

41. DIFFERENTIAL PULLEY

Mechanical Advantage = $\frac{2C}{C - c}$

C is circumference of large wheel; c is circumference of small wheel

42. TEMPERATURE CONVERSION

(1) Centigrade to Fahrenheit
$t_F = 9/5t_C + 32$

t_F is Fahrenheit temperature; t_C is Centigrade temperature

(2) Fahrenheit to Centigrade
$t_C = 5/9(t_F - 32)$

t_C is Centigrade temperature; t_F is Fahrenheit temperature

43. LINEAR EXPANSION

$\Delta l = al(t - t_o)$

Δl is increase in length; a is coefficient of linear expansion; l is original length; t is final temperature; t_o is original temperature

44. KELVIN TEMPERATURE

$T = t_C + 273$

T is Kelvin temperature; t_C is Centigrade temperature

45. CHARLES' LAW

$$\frac{V}{V'} = \frac{T}{T'}$$

V is original volume; V' is new volume; T is original Kelvin temperature; T' is final Kelvin temperature

46. BOYLE'S AND CHARLES' LAWS COMBINED

$$\frac{pV}{T} = \frac{p'V'}{T'}$$

p is original pressure; V is original volume; T is original Kelvin temperature; p' is final pressure; V' is final volume; T' is final Kelvin temperature

47. HEAT EXCHANGE

$Q = mc\Delta t$
Q is amount of heat required; m is mass of substance; c is specific heat of substance; Δt is change in temperature

48. SOUND WAVE FORMULA

$v = f\lambda$
v is velocity of wave; f is frequency; λ is wavelength

49. INTENSITY OF SOUND

$$B = 10 \log \frac{I}{I_o}$$

B is intensity of sound wave; I is energy of threshold of hearing, 10^{-16} watt/cm^2

50. RESONANCE IN TUBES

(1) Closed Tube

$\lambda = 4(l + 0.4d)$
λ is wavelength; l is length of closed tube; d is diameter of tube

(2) Open Tube

$\lambda = 2(l + 0.8d)$
λ is wavelength; l is length of open tube; d is diameter of tube

51. ILLUMINATION

$$E = \frac{I}{R^2}$$

E is illumination; I is intensity; R is distance from source to illuminated surface

52. IMAGES IN MIRRORS AND LENSES

$S_o : S_i = D_o : D_i$

S_o is object size; S_i is image size; D_o is object distance; D_i is image distance

53. IMAGES IN MIRRORS AND LENSES

$$\frac{1}{D_o} + \frac{1}{D_i} = \frac{1}{f}$$

D_o is object distance; D_i is image distance; f is focal length

54. INDEX OF REFRACTION

$$u = \frac{\sin \theta i}{\sin \theta r}$$

u is index of refraction; θ_i is angle of incidence; θ_r is angle of refraction

55. LIGHT WAVE FORMULA

$C = f\lambda$

c is velocity of light; f is frequency; λ is wavelength

56. SIMPLE MAGNIFIER

$$\text{Magnifying Power} = \frac{25\ \text{cm}}{f\ \text{cm}} = \frac{10\ \text{in}}{f\ \text{in}}$$

f is focal length of lens in units indicated

57. COMPOUND MICROSCOPE

$$\text{Magnifying Power} = \frac{25L}{f_e f_o}$$

L is length of tube in centimeters; f_e is focal length of eyepiece in centimeters; f_o is focal length of objective in centimeters

58. REFRACTING TELESCOPE

$$\text{Magnifying Power} = \frac{f_o}{f_e}$$

f_o is focal length of objective; f_e is focal length of eyepiece

59. OHM'S LAW

$$I = \frac{V}{R}$$

I is current; V is potential difference; R is resistance

60. LAWS OF RESISTANCE

$$R = \frac{pl}{d^2}$$

R is resistance; p is resistivity; l is length in feet, d is diameter in mils

61. RESISTANCES IN SERIES

$R = R_1 + R_2 + R_3$ ......

R is total resistance; R_1, R_2, R_3, etc. are individual resistances

62. RESISTANCES IN PARALLEL

$\frac{1}{R} = \frac{1}{R_1} + \frac{1}{R_2} + \frac{1}{R_3}$ .....

R is joint resistance; R_1, R_2, R_3, etc. are individual resistances

63. QUANTITY OF ELECTRIC CHARGE

$Q = It$

Q is amount of charge; I is current; t is time

64. ELECTRIC POWER

$P = IV$

P is power; I is current; V is potential difference

65. ELECTRIC POWER

$P = I^2R$

P is power; I is current; R is resistance

66. LAWS OF ELECTROLYSIS

$m = zIt$

m is mass; z is electrochemical equivalent; I is current; t is time

67. CELL FORMULA

$$I = \frac{E}{R_e + R_i}$$

I is current; E is emf of cell; R_e is external resistance; R_i is internal resistance

68. CELLS IN SERIES

$$I = \frac{nE}{R_e + nR_i}$$

I is current; n is number of cells; E is emf of one cell; R_e is external resistance; R_i is internal resistance

69. CELLS IN PARALLEL

$$I = \frac{E}{R_e + \frac{R_i}{n}}$$

I is current; E is emf of one cell; R_e is external resistance; R_i is internal resistance; n is number of cells

70. JOULE'S LAWS

$$W = I^2Rt$$

W is energy; I is current strength; R is resistance; t is time

71. CAPACITIVE REACTANCE

$$X_c = \frac{159,000}{fc}$$

X_c is capacitive reactance in ohms; f is frequency in cycles; c is capacitance in microfarads

72. INDUCTIVE REACTANCE

$$X_l = 2\pi fL$$

X_l is inductive reactance in ohms; f is frequency in cycles; L is inductance in henrys

73. IMPEDANCE

$$Z = \sqrt{R^2 + (X_l - X_c)^2}$$

Z is impedance; R is resistance; X_l is inductive reactance; X_c is capacitive reactance

74. OHM'S LAW FOR A.C. CIRCUITS

$$I = \frac{V}{Z}$$

I is current; V is voltage; Z is impedance

75. RANGE OF TELEVISION STATION

$$D = 1.23\sqrt{H}$$

D is range in miles; H is height of antenna in feet

GLOSSARY – THE LANGUAGE OF PHYSICS

CONTENTS

GLOSSARY – THE LANGUAGE OF PHYSICS

Aberration, chromatic — the formation of colored fringes on lens images due to the failure of the lens to focus light of all wave lengths at an equal distance from the lens.

Aberration, spherical — a defect of lenses which causes light near the edge of the lens to focus closer to the lens than that falling near the center.

Absolute humidity — the actual quantity of water vapor per unit volume of air.

Absolute pressure — actual pressure, not reduced by atmospheric pressure of 15 lb/in^2.

Absolute temperature — temperature reckoned from absolute zero.

Absolute zero — the temperature at which transmissible motion of particles (heat) ceases.

Absorption — the conversion of radiant energy into energy of a different form, *e.g.*, the sun's energy radiated to the earth is absorbed (partially) and transformed into heat.

Acceleration — rate of change of velocity.

Acceleration, gravitational — the gain in velocity of a body falling freely in a vacuum; 32.08 ft/sec^2 at the equator, 32.16 ft/sec^2 at New York, 32.258 ft/sec^2 at the poles. (See *g*.)

Accommodation — the ability of the eye to adjust its focusing power to the distance of the object viewed.

Achromatic — free from color; having no color fringe. A term applied to lenses.

Acoustics — the study of sound and of hearing. The factors that determine how well an auditorium permits hearing a program on its stage.

Adhesion — the force holding unlike molecules together.

Aerial — the wire, or other conductor that receives radio waves, or that radiates them from a broadcast station.

Aerodynamics — the study of the effect of air acting on moving surfaces as the wing, propeller, or rudder of a plane.

Agonic line — a line connecting points having no compass declination. A compass needle at any point on this line points due north.

Aileron — a hinged flap on an airplane wing used in turning the plane.

Air brakes — brakes operated by compressed air.

Air compressors — pumps used to raise air pressure above normal.

Air conditioning — improving the temperature, the humidity and cleanliness of air.

Air pump — pumps used either to increase or reduce air pressure.

Air speed — the speed of a plane relative to the air in which it moves.

Airfoil — any surface upon which moving air acts to change the motion of an airplane; the wings, flaps, rudder, elevator, and propellers.

Airplane — any heavier-than-air craft that depends upon relative motion of air for its support in flight.

Alpha rays — helium ions, one of the particles emitted by radium.

Alternating current — an electrical current in which the flow of electrons reverses periodically.

Alternator — an A C generator.

Altimeter — an instrument which indicates altitude above a given level.

Amalgamation — combining with or covering a metal with mercury.

Ammeter — an instrument for measuring intensity or rate of transfer of electricity in a circuit.

Ampere — the unit of current intensity, or rate of transfer of electric charge equivalent to 1 coulomb/sec. Legally, the current that will deposit 0.001118 gram of silver per second.

Ampere-turns — the product of the current in amperes × the turns of wire in an electromagnet; this product determines the magnetizing force of the coil.

Amplification — the increasing of the energy of a circuit either in radio or audio frequencies.

Amplifier — a device, usually a radio tube, for increasing the energy of a circuit.

Amplitude — the amount of displacement in any oscillatory motion; usually applied to the motion of an air particle displaced by a sound wave, or to the motion of a vibrating string or pendulum.

Aneroid barometer — an easily portable gage for indicating atmospheric pressures; it is *without liquid* — the earlier barometers having been made with mercury.

Angle of attack — the angle between the wing of a plane and the line of its motion relative to air.

Angle of incidence — the angle between a light ray falling on a surface and a line normal (perpendicular) to that surface. (See also *angle of reflection.*)

Angstrom — a small unit of length usually associated with the measurement of light, or other, waves. $1\ \text{A}° = 10^{-8}$ cm.

Anode — the positive electrode of any electrochemical apparatus.

Antinode — the part of a vibrating body that has the greatest amplitude.

Arc — the luminous curved discharge between oppositely charged terminals, as in an arc lamp.

Archimedes' principle — the buoyant force on a submerged body equals the weight of the displaced fluid.

Armature — a wire-wound core of iron acting as the rotor of a dynamo; more loosely, any piece of iron in which electromagnetic effects are utilized or produced.

Astigmatism — a defect of the eye, or other lens indicated by a failure to focus equally in all planes.

Atmosphere — the gaseous envelope or portion of the earth. Also used to designate normal atmospheric pressure, or 14.7 lb/in^2.

Atom — the smallest particle of an element.

Atomic bomb — an explosive that utilizes the energy liberated by the breaking up of a heavy atom into smaller atoms.

Atomic number — the number of protons in the nucleus of an atom.

Audio-frequencies — vibrations whose frequencies lie within the normal hearing range.

Audio signal — any information received by a radio set.

Back emf — a counter-voltage induced in the armature coils of a motor.

Bar — a unit for stating pressure (usually atmospheric pressure); one bar is equivalent to a pressure of 1 million dynes/cm^2. (See *millibar.*)

Barograph — a self-recording barometer.

Barometer — any pressure gage for measuring atmospheric pressure.

Battery — any combination of electric cells; incorrectly, a single cell.

Beats — rhythmic changes in loudness of the combined sounds from two sources of slightly different frequencies.

Beat frequency — a term used generally in radio to denote the *difference* in frequency of two oscillating circuits.

Bel — a logarithmic expression of the ratio of two quantities as, for example, the loudness of two sounds. (See *decibel.*)

Bernouilli's principle — at a constant level the sum of the kinetic and potential energies of a fluid is constant, meaning that if its velocity increases its pressure must decrease and vice versa.

Beta rays — electrons emitted by radioactive materials.

Bimetal — a bar of two unlike metal strips fastened together which curves when heated due to the unequal expansion of the two metals. (See *compound bar* and *thermostat.*)

Block and tackle — any combination of fixed and movable pulleys.

Boiling — changing a liquid to a vapor whose pressure equals the (atmospheric) pressure on the liquid.

Boiling point — the temperature at which the vapor pressure of a liquid equals the pressure on the liquid.

Bolometer — a device for measuring temperature by utilizing the very constant relation between temperature and electrical resistance.

Bourdon gage — a common form of pressure gage.

Boyle's law — the product of the volume of a mass of gas and its pressure is a constant, meaning that as volume increases pressure decreases, and conversely.

British Thermal Unit (B T U) — the heat required to raise the temperature of 1 pound of water by 1 degree Fahrenheit.

Brownian movement — irregular microscopic movements of particles of a solid suspended in a fluid caused by being struck by molecules of the fluid.

Brush — a piece of carbon or metal used on motors and generators to make sliding contact between a stationary circuit and a rotating circuit.

Buoyancy — the lifting effect of a fluid upon a body immersed in it.

Caisson — a water-tight chamber in which workers may do underwater construction work under pressure.

Calorie — the heat required to raise the temperature of 1 gram of water 1 degree centigrade. (Also used in dietetics for one thousand times this much heat.)

Cam — a mechanical device commonly used to transform rotary motion into back-and-forth motion.

Camera — an optical device in which a convex lens system forms a permanent image on sensitized film.

Camshaft — a rotating shaft upon which cams are mounted.

Candle (candle power) — the unit of light intensity equivalent to 1 lumen per unit solid angle.

Candle, standard — the candle designated by the U. S. Bureau of Standards as one whose rate of light emission is one candle power. (A more reliable standard has superseded the candle.)

Capacitance — the ratio between the charge on either plate of a capacitor (condenser) and the emf charging it; *i.e.*, how much charge can be stored at a given voltage.

$$\text{Capacitance in farads} = \frac{\text{coulombs}}{\text{volts}}.$$

Capacitor — or *condenser*, an insulator whose opposite sides, covered by conducting plates, receive and store opposite electrical charges.

Capacity reactance — the measure in ohms of the effect of introducing a capacitance into an oscillating circuit.

Capillary action — a general term for the behavior of a liquid whose surface is in contact with a solid such as a tube or other narrow space.

Carburetor — a gas-engine device for vaporizing liquid fuel (gasoline), and mixing it with air.

Carrier wave — the steady wave from a radio transmitter upon which the modulations of the program wave are superimposed.

Cathode — (*a*) the negative terminal of devices using electric current. (*b*) the source of electrons in a vacuum tube.

Cathode rays — electrons shot off from a highly-charged negative body.

Cell — in electricity, a producer of emf as the result of chemical action, light, or heat.

Center of gravity — the point about which all the gravitational moments are in equilibrium; *i.e.*, the point at which the body can be balanced by a single upward force.

Centigrade scale — a thermometer scale having one hundred gradations between its zero (freezing point of water) and the boiling point of water at normal pressure.

Centrifugal force — in rotary motion, the reaction to the force that keeps the body in a circular path.

Centrifugal pump — a common and useful pump in which rotating blades throw the fluid outward into a space leading to the discharge pipe.

Centripetal force — the force needed to cause a body to travel a circular path rather than a straight path.

Cgs system — a metric system of measurement in which the three fundamental units are the *centimeter*, *gram*, and *second*.

Chain reaction — a series of events in which energy or materials liberated in the first step brings about the next, and succeeding, steps.

Change of state — the change in physical condition undergone by a substance that is melting, vaporizing, condensing, freezing, or subliming. Heat changes occur during all changes of state.

Charge — the condition brought about by adding or removing electrons from a neutral body.

Charles's law — the volume of a given mass of gas varies directly as the absolute temperature.

Choke coil — a coil, usually an electromagnet, whose inductive reactance reduces the working voltage of an A C circuit.

Chord — in music, combinations of notes whose frequency ratios are expressed in small whole numbers, *e.g.*, 4:5:6.

Chromatic scale — a musical scale consisting of eight full tones and five semitones (sharps and flats).

Circuit, electrical — a complete path in which an electric current may exist.

Circuit breaker — an electromagnetic device used to break a circuit automatically if the circuit is overloaded.

Circular mil — a unit for expressing the cross-section area of a wire, found by squaring the diameter of the wire in thousandths of an inch.

Cloud track chamber — a space in which electric charges betray their motion by condensing vapor along their paths.

Coefficient of expansion — a constant that shows the linear — or volumetric — expansion of one unit of a given material for one degree rise in temperature.

Coefficient of friction — the ratio between the force needed to move a body along a surface and the force holding the body against that surface; in a simple case, pulling force divided by weight of body pulled along a level surface.

Cohesion — the force of attraction between like molecules.

Coil, induction — an open-core transformer used to boost the voltage of low-voltage D C sources.

Color — one aspect or characteristic of a visual sensation which depends upon the wave length of the radiation received by the eye.

Commutator — a metallic ring split into two or many segments used on a generator to convert an internal A C to an external D C: used on motors to reverse the armature polarity each time a field pole is passed. (See *motor* and *generator*.)

Compass, earth inductor — a direction indicator which depends upon the rate at which the earth's magnetic field is cut by a revolving coil carried by an airplane.

Compass, gyroscopic — a direction indicator depending upon the tendency of the axis of a heavy rotating wheel to point north-south.

Compass, magnetic — a pivoted magnet used, with suitable charts, to determine direction.

Complementary colors — any two colors that may combine to produce the effect of white light.

Component of a force — the part of a given force that acts effectively in a given direction.

Composition of forces — a method of finding a single force (resultant) that will replace two or more given forces.

Compound bar — two thin flat strips of different metals arranged so that their unequal expansions may be used in thermometers or thermostats. (See *bimetal*.)

Compound-wound generator — a D C generator whose field magnet is excited by both a shunt coil and a series coil.

Concave lens — any lens thinner in the middle than at the edges.

Concave mirror — a mirror whose center curves away from the observer.

Condensation — (*a*) in sound: that part of the sound wave in which the medium (usually air) is compressed; (*b*) the changing of a vapor into its liquid.

Condenser — (*a*) electrical, see *Capacitor;* (*b*) the part of a distillation apparatus in which the pure liquid is formed from its vapor; (*c*) a compartment in which exhaust steam is reconverted into water.

Conductance — the reciprocal of electrical resistance; the ratio of current to potential difference.

Conduction — the transmission of heat or of electrical energy by the motion of the particles of the conductor.

Conductivity — the ability to transmit electrical or thermal energy through the particles of the conductor.

Conductor — a substance able to transmit heat or electricity.

Conservation of energy — energy can neither be created nor destroyed (first law of thermodynamics).

Conservation of matter — in the ordinary sense, matter can neither be created nor destroyed, but under special conditions it may apparently be transformed into radiation energy.

Conservation of momentum — momentum, like energy and matter, can neither be created nor destroyed but can only be transferred from one body to another.

Convection — the transfer of heat by currents in fluids.

Conversion of energy — the disappearance of energy in one form and its reappearance in another; *e.g.*, heat may be converted into electrical energy.

Converter — a combined motor and generator used to convert A C to D C or the reverse.

Convex lens — a lens thicker in the middle than at its edges.

Convex mirror — a mirror whose center curves toward the observer.

Coolidge tube — the common form of X-ray tube.

Coulomb — (*a*) the unit of electrical charge, equivalent to the loss or gain of 6.3×10^{18} electrons. (*b*) a French scientist.

Coulomb's laws — (*a*) for electric charges: the force between two small charges depends upon their product divided by the square of the distance between them; (*b*) for magnetic poles: the force between two magnetic poles depends upon the product of the pole strengths divided by the square of the distance between them.

Couple — a pair of forces, both tending to produce rotary motion in the same direction.

Critical angle — the largest angle at which a ray of light may strike a rarer medium and still be refracted into the rare medium.

Critical pressure — the pressure of a gas at its critical temperature.

Critical temperature — the temperature at which a liquid vaporizes regardless of pressure.

Crookes tubes — general name for high-vacuum tubes in which electrical discharge is brought about through ions of the remaining gas.

Crystal — (*a*) in general, a solid whose particles are arranged in an orderly pattern. (*b*) in radio and telephony, a piece of a quartz (or other material) crystal so cut as to respond electrically to mechanical pressures and vice versa.

Current — (*a*) any continuous transfer of electric charges. (*b*) the motion of a fluid in a given direction, as in a convection current.

Cycle — (*a*) in general, a repeated series of events. (*b*) in A C circuits, the changes in voltage passed through before any given voltage is repeated. (*c*) in engines, the operations performed before any given operation is repeated.

Cyclone — a large storm which develops spirally around low-pressure areas in temperate regions usually attended by higher temperatures, moderate winds, and steady precipitation.

Cyclotron — a large and complicated apparatus for speeding up sub-atomic particles until atoms may be broken down by their impact.

Cylinder — the expansion chamber of a reciprocating steam engine; the combustion chamber of a gasoline or diesel engine; the working chamber of various pumps.

D'Arsonval galvanometer — a galvanometer whose moving element is a coil suspended or pivoted between the poles of a magnet; the original form of many present-day electric meters.

Decibel — a unit for comparing the intensity of a sound wave with another chosen as a standard.

Declination — the angle between the magnetic compass and true north.

Density — the mass per unit volume of a substance. (See *optical density.*)

Detector — a device, usually a triode, for rectifying current oscillations so that they can affect a speaker.

Deuteron — the nucleus of "heavy" hydrogen.

Deviation of a compass — an error in compass behavior due to local conditions, as the iron of a ship.

Dew point — the temperature at which atmospheric water vapor condenses.

Dielectric — an electrical non-conductor.

Dielectric constant — the factor in the capacitance of a capacitor that depends upon the material of the insulator (dielectric) used.

Diesel engine — an internal-combustion engine whose fuel oil is ignited by the high compression of the intake air in the cylinder.

Diffusion — (*a*) the mixing of gases or liquids of different densities by the individual motions of their molecules. (*b*) the scattering of light which occurs when light falls upon rough reflectors.

Diode — a vacuum tube of two elements — plate and hot filament.

Dip — or *inclination*, the angle between a dipping needle (pivoted to swing in a vertical plane) and the horizontal.

Direct current — an electric current flowing in one direction only.

Dirigible — a lighter-than-air ship that can be propelled and steered.

Dispersion — the separation of any light into its component wave lengths.

Displacement — the weight of fluid, estimated or actual, that is pushed aside by a body immersed or floating in the fluid. (*Note:* This word has several nautical meanings.)

Dissipation of energy — the transformation of energy into unavailable forms such as low-grade heat.

Dissociation — the separation of an electrolyte into ions.

Distillation — the purification of a liquid or its separation from other liquids by vaporization and subsequent condensation.

Doppler effect — the change in the pitch of sound or the color of light due to the rapid change in the distance between source and observer.

Drag — the opposition offered by the air to the motion of an airplane through it.

Ductility — the property of a substance, usually a metal, that allows it to be drawn into wire.

Dynamics — the part of physics that deals with forces producing motion; observe also aerodynamics, electrodynamics, thermodynamics in which forces and motions are associated with the air, with electricity, and with heat.

Dynamo — usually, a D C *generator* but also the same machine used as a *motor;* hence, a machine for converting mechanical energy into electrical energy or vice versa.

Dyne — the absolute unit of force defined as the force required to give a one-gram mass an acceleration of 1 cm per second.

Eccentric — an off-center disk on the main shaft of an engine, used to obtain back-and-forth motion from the rotary motion of the disk.

Echo — a well-reflected sound.

Eclipse — the darkening of the sun or moon; the solar eclipse occurs when the moon's shadow falls upon the earth; the lunar eclipse occurs when the moon enters the earth's shadow.

Efficiency — the ratio of useful energy delivered (output), to the total energy expended on a machine (input).

Elasticity — the tendency of a deformed body to return to its original shape or size.

Elasticity of length — the force per unit area required to stretch one unit length of a substance to two inches. (Young's modulus.) Elasticity of shape and bulk are also stated by appropriate moduli.

Elastic limit — the point to which a body can be deformed and still return to its original shape when the stress is removed.

Electric, electrical — terms preceded by these adjectives will, where expedient, be alphabetized under their own initial.

Electricity — the general name for all phenomena related to the behavior of charged bodies or particles; a manifestation of energy convertible into other forms of energy.

Electrification — the process of charging a body electrically; the act of adding surplus electrons or of removing normal electrons from a body.

Electrodes — general name for the terminals of any electrical apparatus, particularly apparatus used in electrochemical work.

Electrolysis — the production of a chemical change by electrical means.

Electrolyte — a solution that can conduct an electric current.

Electromagnet — any device for the establishing of a magnetic field by the flow of electric current; in practice, an iron core wound with a coil of wire through which a current is passed.

Electromagnetic induction — the establishing of an emf by the motion of a conductor across a magnetic field.

Electromagnetic waves — transverse waves radiated through space, which are caused by oscillations of electrons or groups of electrons; the general name for the waves which are grouped as radio, infra-red, light, ultra-violet, X-rays and probably cosmic rays.

Electromotive force (emf) — the electrical force existing between two bodies of different potential (charge), and which tends to move electrons so as to equalize this difference. 1 volt of emf is produced when 1 joule of work is done in moving 1 coulomb of charge to a higher potential.

Electron — indivisible particles each bearing a unit negative charge, and found as one of the constituent parts of every atom.

Electron microscope — a device in which an electron beam is focused by electric and magnetic fields so as to form an image on a fluorescent screen.

Electron tubes — vacuum tubes usually containing three elements — cathode, grid, and plate — used to control very small streams of electrons.

Electroplating — the coating of various articles by giving the article such an electric charge as will enable it to attract and hold ions of the plating material.

Electroscope — an instrument for detecting electric charges.

Energy — the ability to do work.

Energy, atomic — the energy possessed by an atom due to the force that holds its particles together (like a tank of compressed air).

Energy, kinetic — energy due to motion.

Energy, potential — energy due to strain.

Engine — a device for converting heat energy into work; more loosely, any energy converter.

Equilibrant — a single force that counterbalances one or more other forces.

Equilibrium — a condition in which neither linear nor angular velocity changes.

Erg — the absolute unit of work (energy) equivalent to the work done by one dyne of force acting through one centimeter of distance; a dyne-centimeter of work.

Evaporation — the changing of a liquid to a gas or vapor.

Exciter — the D C generator whose current supplies the field magnet of an alternator.

Factor of safety — the number of times the necessary strength of a construction is increased to allow for unusual needs.

Fahrenheit scale — a thermometer scale having fixed points of 32° (freezing point of water) and 212° (boiling point at normal pressure).

Fall of potential — the voltage lost when work is done by a charge moving from one point to another; *e.g.*, 4 volts are lost between A and B if 20 joules of work are done when 5 coulombs move from A to B.

Farad — the unit of capacitance equivalent to storing one coulomb at a potential of 1 volt.

Faraday's law — of electrolysis: the amount of any element set free by electrochemical action depends upon current (amperes), time, and the electrochemical equivalent of that element.

Fission, nuclear — the splitting of a heavy atom into two relatively large atoms of different elements.

Fluids — any substance able to flow; the term is applied to include both liquids and gases.

Fluorescence — the simultaneous emission of light waves while absorbing shorter waves (ultra-violet or X-rays).

Flux density — the magnetic strength of an area expressed in lines of force per square centimeter: gausses.

Flux, magnetic — the total number of lines of magnetic induction in any region: maxwells.

Focal length — the distance between the principal focus of a lens (or mirror) and its optical center.

Focus, principal — the point at which a lens converges light waves perpendicular to its principal axis.

Foot-candle — a practical unit of illuminance equivalent to that of a surface 1 foot from a 1 candle-power source.

Foot-pound — the unit of work (energy) equivalent to the work done by a force of 1 pound acting through a distance of 1 foot.

Foot-pound-second system — a system of measurement in which the three fundamental units are the *foot*, *pound*, and *second*. Compare *cgs* and *mks*.

Force — a push or a pull; the work done per unit of distance.

Freezing point — the temperature at which a liquid changes to a solid at normal pressure.

Frequency — the number of oscillations per second made by a sounding body, an alternating current, or a radio transmitter; *wave frequency* is the number of waves per second passing a given point.

Friction — the force that opposes motion between two surfaces in contact.

Front — a weather term describing the boundary between two air masses: usually a region of most rapid weather changes.

Fulcrum — (*a*) an actual point about which a lever rotates; (*b*) any point in a lever about which force moments are calculated.

Fundamental — the lowest note a sounding body can produce and the one produced by the vibration of the body as a whole.

Fuses — strips of metal set into an electric circuit to break the circuit by melting if the circuit is overloaded.

Fusion, nuclear — the union of two, or more, light atoms to make an atom of a different element.

"g" — the symbol for the constant of gravitational acceleration, approximately equivalent to 32 ft/sec^2 or to 980 cm/sec^2.

Galvanometer — a device for detecting and comparing electric currents; the basic instrument which is modified to make both voltmeter and ammeter.

Gas — a state of matter having no definite size or shape but which takes both from its container.

Gauss — the unit of flux density; one line of force per square centimeter.

Geiger counter — a widely used apparatus for detecting the radiations of radioactive elements.

Geissler tubes — low vacuum tubes in which electrical discharges may be studied or utilized.

Gilbert — the magnetomotive force that would establish one line of force in a magnetic circuit having a reluctance of 1 unit.

Gravitation — the force of attraction that appears to exist between all bodies in the universe.

Gravity — the name applied to the apparent force of attraction between the earth and bodies near it.

Grid — the element in a triode whose polarity changes control the flow of electrons from *cathode* to *plate.*

Half-life — the time required for half of the atoms of a given mass of a radioactive element to disintegrate.

Heat — the kinetic energy of molecular motion (plus whatever potential energy is acquired by expansion).

Heat of fusion — the quantity of heat per gram required to change a solid to a liquid without change of temperature.

Heat of vaporization — the quantity of heat per gram required to change a liquid to its vapor without change of temperature.

Heat, specific — the ratio of heat required to raise 1 gram of a substance 1 degree to that required by water for the same change.

Helicopter — an airplane whose large horizontal propeller makes it capable of vertical flight.

Henry — the unit of electrical inductance.

Horsepower — the unit of power equivalent to the doing of 33,000 foot-pounds of work per minute (550 foot-pounds per second).

Humidity, relative — the ratio, in the form of a percentage, of the actual amount of water vapor in the atmosphere to the amount of vapor needed for saturation. (See also *absolute humidity.*)

Hydraulic press — a device by which a small force is multiplied by causing a liquid to exert pressure against a large area.

Hydrometer — a device for measuring the specific gravity of liquids.

Hygrometer — a device for measuring relative humidity.

Illuminance — a measurable quantity that describes the amount of luminous flux received by one unit area of surface, *e.g.*, lumens/meter^2; more informally, a relative quantity expressed in *foot-candles* $= \frac{\text{candles}}{\text{feet}^2}$.

Illumination — a general term describing how well a surface is, or should be, lighted.

Image — the reproduction or counterpart of an object as formed by any optical device.

Impedance — the total opposition offered by all parts of a circuit to the flow of alternating current; the vector sum of resistance, capacitance, and inductance; numerically, the ratio of impressed voltage to current intensity.

Impenetrability — the property of being unable to occupy space taken by another body.

Impulse — the product of a *force* and the *time* during which it operates; impulse determines momentum.

Index of refraction — (*a*) the ratio of velocity of light in vacuum to its velocity in a given medium; (*b*) the ratio of the sine of the angle of incidence (in a vacuum) to the sine of the refraction angle in a given medium.

Inductance — the measure of the ability of a coil to self-induce a counter-voltage depending upon the rate at which the current varies.

Induction — *magnetic,* the production of a magnet by use of the field of another magnet. *Electrostatic,* the charging of one body by use of the electric field of another charged body.

Inertia — the opposition offered by a body to any attempt to change its rate or direction of motion.

Infrared waves — the waves in the electromagnetic spectrum whose lengths are from 0.3 mm to 0.00075 mm.

Intensity — a term used in describing quantities related to (*a*) electric currents; (*b*) electric charges; (*c*) light sources; (*d*) magnets and magnet fields; (*e*) sound waves. Its meaning varies with its usage.

Interference — *in general,* the mutual effect of two waves (sound, light, or others) that arrive at the same point at the same time; *restricted meaning,* the annulment of two equal wave trains when they arrive out of phase.

Ionization of gases — the separation of gas atoms into charged particles, usually as the result of forcible collision with other particles.

Ions — charged particles in a fluid; ions may be atoms or groups of atoms which have gained or lost electrons, or, in a gas, may be free electrons.

Iso- — a prefix meaning *equal;* hence *isobar*, a line drawn through places of equal pressure; *isocline* — through places of equal dip; *isogon* — through places of equal compass declination; *isotherm* — through places of equal temperature.

Isotopes — forms of the same element which have the same number of planetary or ring electrons but which have different atomic masses (number of neutrons).

Jet engine — one of several types of heat engine in which *thrust* is obtained directly from the action of high-temperature gas particles colliding with fixed parts of the engine itself.

Joule — a unit of work(*a*) in electrical systems it is the equivalent of moving 1 coulomb through a potential difference of 1 volt; (*b*) in mechanical systems it is the work done by 1 newton of force acting through 1 meter of distance.

Kelvin scale — the temperature scale in which the zero point is *absolute* zero; the freezing point of water is 273°; boiling point of water is 373°, etc. More commonly called the *absolute* scale.

Kilo- — a prefix meaning *1000 of;* hence a *kiloampere* is 1000 amperes; used similarly in *kilocycle* — 1000 oscillations per second; *kilometer, kilovolt, kilovoltampere, kilowatt, kilowatt-hour.*

Kinescope — the modified cathode-ray oscilloscope that forms the television picture that you see.

Lenses — pieces of transparent material having at least one curved surface which makes it able to refract light so as to make or help to make an image.

Lenz's law — an induced current will set up a magnetic field that will oppose the motion required to induce the current.

Lever — in practice, any device upon which an effort is applied so as to rotate the system about a fixed point against an opposing force.

Leyden jar — an early type of capacitor.

Lift — the perpendicular component of the force of the air against an airplane; the component that is effective in supporting the plane's weight.

Light — that part of the electromagnetic spectrum (wave lengths between 0.00081 mm and 0.00036 mm) which is capable of producing vision.

Liquid — a state of matter having definite volume (at constant temperature), but which takes the shape of its container.

Loran — a navigational system of finding direction by comparison of the arrival times of signals from two widely-separated radio stations.

Loudness — one of three characteristics of a sound — or aspects of an auditory sensation — depending primarily upon the amplitude of the sound wave.

Lumen — the unit of luminous flux equivalent to the amount of flux intercepted by 1 square foot of surface held perpendicularly at a distance of 1 foot from a standard candle (or on 1 square meter at a distance of 1 meter). A standard candle emits luminous flux at the rate of 12.57 ($4\ \pi R^2$) lumens.

Luminous body — one which emits waves of visible light.

Machine — a device to overcome a large resistance with a small force.

Magnetic circuit — the complete path of the magnetic lines of force (flux).

Magnetic field — space surrounding a magnet in which magnetic force is effective.

Magnetic pole — a part of a magnet at which the attraction for magnetic substances is greatest.

Magneto — an electric generator the field of which is produced by a permanent magnet.

Magnetomotive force — the force which sets up magnetic flux.

Malleability — the ability of a material to be hammered or rolled into sheets.

Manometer — a gage for measuring fluid pressure.

Mass — quantity of matter in a body; a measure of its inertia.

Mass spectrograph — an apparatus in which electrical and magnetic fields are applied to rapidly moving particles so as to (*a*) determine the nature of the particles by the deflection produced, and (*b*) to separate the particles according to their masses.

Matter — the material of which substances are made, and on which energy acts.

Maxwell — one line of magnetic flux.

Mechanical advantage — (*a*) *ideal:* the ratio of effort-displacement to resistance-displacement in a machine; (*b*) *actual:* the ratio of the resistance to the effort force that overcomes it.

Mechanical equivalent of heat — the amount of work required to produce a unit quantity of heat: 778 ft-lb = 1 B T U; 4.18 joules = 1 calorie.

Mechanics — the study of the effect of forces on bodies.

Medium — substance through which waves travel.

Melting point — temperature at which a solid changes to a liquid.

Meter — unit of length in the metric system = 39.37 in.

Mho — unit of electrical conductance.

Microphone — a device for changing sound waves into varying electric currents.

Microscope — an optical instrument for making magnified images of small objects.

Mil — a thousandth of an inch; used in calculating the resistance of wires.

Mil-foot — a round wire 1 foot long and 1 circular mil in area.

Millibar — a unit of atmospheric pressure; equivalent to one thousand dynes per square centimeter.

Mirror — a surface that forms an image by the reflection of light.

Mks system — a system of measurement whose fundamental units are the *meter*, *kilogram*, and *second*. (See *cgs* and *ft-lb-sec.*)

Modulation — altering continuous waves by means of currents from a varying microphone receiving sound waves.

Molecule — smallest particle of a substance which possesses the properties of the substance.

Moment of a force — its turning effect, measured by the force times its perpendicular distance from the fulcrum.

Momentum — quantity of motion, measured by mass times velocity.

Motion — change of position.

Motor-generator — combination of a motor driving a generator mounted on the same shaft.

Multiplier phototube — a sensitive device for the detection of radioactivity.

Negatively charged body — one that has an excess of electrons.

Neptunium — an element produced by neutron bombardment of uranium. It transmutes into plutonium.

Neutron — an uncharged particle in the nucleus of an atom.

Node — a point of no vibration, as in a string.

Newton — the unit of force in the mks system: the force that will, if applied to a 1 kilogram mass, produce an acceleration of 1 meter/sec^2.

Nucleus — the positively charged core of an atom.

Ohm — unit of electrical resistance.

Ohm's law — the current in an electrical circuit in amperes equals the electromotive force in volts divided by the resistance in ohms.

Omnirange — a navigational system by which a plane determines its position and course by reference to radio beams from many stations.

Opaque bodies — bodies that stop the passage of light.

Optical center — of a lens, a point so located that rays passing through this point will not bend.

Optical density — opposition of a transparent substance to the passage of light; its measure is the index of refraction.

Organ — a musical instrument producing tones by the vibrations of air in pipes, by reeds, or by electrical impulses.

Orthicon — the apparatus that scans a scene for television and sets up the currents and hence the waves that transmit the video signals to your antenna.

Oscillations — (*a*) back-and-forth motions as of a pendulum, or (*b*) of electrons in a radio circuit.

Oscillator — any device or circuit that emits waves: the waves may be sonic, supersonic, or electromagnetic.

Oscillograph — an apparatus for converting oscillations of a circuit into visible curves, scenes, or other signals.

Overtones — tones produced by a body vibrating in segments.

Parallel circuit — an electric circuit in which the current is divided between two or more conductors.

Pascal's law — pressure applied to an enclosed fluid is transmitted equally in all directions and acts with equal force on equal areas.

Pendulum — a swinging mass, vibrating in equal times.

Penstock — a pipe which carries water to a turbine.

Pentode — an electron tube containing a filament, plate and three grids.

Penumbra — partially illuminated part of a shadow.

Permalloy — one of several easily magnetized alloys of iron and nickel.

Permeability — ability of a substance to concentrate lines of magnetic force.

Phosphorescent bodies — bodies which emit light after having been illuminated.

Photoelectric cell — a device for causing the release of electrons when acted on by light.

Photometer — an instrument for measuring the intensity of a light source.

Photon — the energy packet, or *quantum*, associated with a train of light waves or other electromagnetic waves.

Piezo-electric effect — the conversion of pressure changes by certain crystals into electrical impulses.

Pigment — material used to give color to a paint.

Pile, atomic — an assembly of graphite blocks in which uranium is transformed to plutonium.

Pitch (screw) — distance between the threads of a screw.

Pitch (sound) — characteristic of a sound which determines its place on the musical scale.

Plane wave — one whose wave-front is a plane surface.

Plano-lenses — lenses flat on one side.

Plastic substance — one which remains deformed after an applied force ceases to act. (Commonly used now to describe substances easily molded, especially when heated.)

Plate — the electrode in a vacuum tube which receives electrons from the cathode and delivers them to an external circuit.

Plutonium — an element produced by transmutation of neptunium. The material of some atomic bombs.

Pneumatic tool — a tool operated by compressed air.

Polar front — the face of a mass of cold air in the atmosphere.

Polarization — (*a*) of a cell, deposit of hydrogen on the positive plate reducing the voltage and the current; (*b*) of light, the reduction of ordinary light to waves whose vibrations are in one plane only.

Polarized light — light whose vibrations are in one plane only.

Polar mass — a large body of cold air, originating in the north in this hemisphere.

Pole pieces — projecting ends of electromagnet cores, particularly in motors and generators.

Positively charged body — one that has lost electrons.

Potential — in electricity: the ratio between the potential energy possessed by an electric charge and the amount of the charge itself.

$$\frac{\text{Potential}}{\text{(in volts)}} = \frac{\text{joules (of potential energy)}}{\text{coulombs (of electric charge)}}.$$

Potential drop — loss of voltage as a current is forced through a resistance.

Pound — the unit of force in the ft-lb-sec system; it gives a mass of 1 slug an acceleration 1 ft./sec^2.

Poundal — unit of force, about 1/32 of a pound.

Power — time rate of doing work.

Power-amplifier — a radio circuit which enormously increases the feeble impulses received from a crystal oscillator.

Power factor — ratio of true power (watts) to apparent power (volt-amperes) in an A C circuit.

Precipitation — condensation of water vapor in the atmosphere as rain, snow, or hail.

Pressure — force exerted on a unit area.

Primary colors — red, green, and blue, which will combine to form white light.

Primary (transformer) — the coils through which the current entering the transformer pass.

Principal axis — (*a*) of a lens: a line passing through the centers of curvature; (*b*) of a mirror: a line through the center of curvature and the mid-point (vertex) of the curved mirror.

Principal focus — point to which parallel rays of light are brought by a lens or mirror.

Prism — a transparent body whose polished surfaces are at an angle.

Projector — an optical instrument for throwing pictures on a screen.

Propeller — the set of rotating airfoils which drives or pulls an airplane through the air.

Proton — a positively charged particle in the nucleus of an atom.

Pulley — a wheel over which a rope passes, used singly or combined with others as a machine.

Pump — a machine for producing a flow of liquid or gas.

Quality (sound) — one of the three characteristics of a sound that enables the ear to distinguish the notes of one musical instrument from another even if the sounds are alike in pitch and loudness. (*Note:* there is no synonym or exact definition for aspects of sensations such as sweetness, color, pitch, etc.)

Quantum — the amount of energy associated with electromagnetic waves of a given frequency. (See *photo.*)

Quantum theory — a theory that radiant energy is emitted or absorbed only in definite amounts (*quanta*).

Radar — a device for locating distant objects by the reflection of very short radio waves.

Radiation — (*a*) the transfer of energy by electromagnetic waves; (*b*) the emission of particles and/or waves by radioactive bodies.

Radiator — part of the cooling system of an automobile.

Radio — a general term covering the transmission of audible signals or programs by electromagnetic waves.

Radioactive elements — elements that radiate waves and/or particles either naturally or after having been made artificially radioactive.

Radio-frequencies — frequencies above that of audible sound, from 15,000 up to billions of cycles.

Radio-frequency amplification — amplification of radio oscillations before they reach the detecting tube.

Radioisotope — any isotope capable of emitting waves and/or particles until it reaches a steady state.

Radio waves — the electromagnetic waves, caused by the oscillation of electric charges, and converted into sound by receiving sets.

Radium — the best known radioactive element.

Rarefaction — the part of a sound wave in which the particles are moving away from each other.

Ray — a narrow portion of a light stream.

Reactance — opposition to the flow of A C produced by inductance or capacitance.

Reaction — one of a pair of equal and opposite effects produced by the action of a force.

Real image — one from which light actually comes to the eye.

Receiver — part of a telephone or radio set by which varying electric currents are converted into sound waves.

Rectifier, power — a device for changing A C into D C.

Rectifier, radio — a device for converting alternating current waves into impulses in one direction.

Reflection — bending back of waves.

Refraction — bending of waves on passing from one medium into another.

Regenerative circuit — a radio circuit in which current in the grid circuit is increased by inductive action in the plate circuit.

Regular reflection — the reflection of undistorted light waves, which may result in the formation of an image of the source of light. Echoes result from regularly-reflected sound waves.

Relative humidity — the ratio (usually in the form of per cent) of the weight of water vapor in a given volume of air to the weight of vapor required for saturation at the same temperature.

Relay — a switch operated by a small electric current to control a larger current.

Reluctance — opposition of a substance to becoming magnetized.

Remote control — control of distant electric machinery by a relay system.

Residual magnetism — magnetism remaining in a magnet core after the magnetizing force ceases to exist.

Resistance, electrical — the property of an electrical conductor that determines the current (amperes) between two points of different potential (volts). Numerically: the ratio of volts to amperes in a conductor.

Resistance, machines — the force exerted *by* a machine.

Resistance box — a set of coils of known resistance.

Resolution of forces — finding the components of a force which act in specified directions.

Resonance — any response made by one oscillating system to impulses from another system.

Resonant circuit — one in which the inductive and capacitive reactances are equal.

Resultant force — a single force equivalent to the combined effect of two or more given forces.

Retina — portion of the eye sensitive to light.

Reverberations — repeated reflections of sound.

Rheostat — a device for regulating electric current by resistance.

Rotary motion — circular motion. This change of position of a rotating body involves such quantities as angular acceleration, displacement, momentum, speed, and velocity, all of which correspond to the linear quantities treated in this book.

Rotor — the rotating part of such machines as motors and generators, especially those used in A C.

Saturation — (*a*) condition of air when it has all the water vapor it can hold. (*b*) condition of a magnetic material when it has all the lines of force it can hold.

Secondary, transformer — the coils through which the current delivered by a transformer pass.

Segments — (*a*) one of the parts in which a body vibrates. (*b*) one section of a commutator.

Selective reflection — reflection of one color more than others.

Self-induction — inductive effect of a coil which cuts its own magnetic field as current varies.

Series circuit — one in which the same current passes through two or more resistances, one after the other.

Series-wound — motors or generators whose armature and field coils are in series.

Sextant — an optical instrument used by mariners and others to determine the angular distance between two bodies, usually the sun (or a star) and the horizon.

Shadow — space from which light is cut off by an opaque object.

Shunt circuit — one in which the electric current is divided between two paths.

Shunt-wound — motors or generators whose armature and field coils are in shunt.

Siphon — a bent tube used to convey a fluid over a small elevation.

Siren — an instrument for producing sound by a succession of puffs of air.

Slip rings — metallic rings by which the armature of an A C dynamo is connected to the brushes.

Slug — the unit of mass in the ft-lb-sec system equivalent to the weight of the body in pounds divided by the value of "g" at the site where the observation is made. (*Note:* This usage has been adopted to avoid the use of the unit *pound* for both force and mass.)

Solar — related to the sun.

Solenoid — coil of wire for an electromagnet.

Solution — a uniform mixture in a liquid state.

Sonar — a device to locate objects under water by the reflection of sound waves.

Sonometer — an instrument for studying the vibrations of a musical string.

Sound pictures — moving pictures accompanied by sound.

Sound track — a narrow strip of dark and light bands alongside the pictures by which sound is recorded on a movie film.

Specific gravity — the ratio of the weight of a substance to the weight of an equal volume of water.

Specific heat — ratio of heat required to raise 1 gram of a substance 1 degree C to that required by 1 gram of water for 1 degree C temperature rise.

Specific resistance — resistance of a conductor of unit length and unit cross-section (usually 1 mil-foot).

Spectroscope — an instrument for forming and studying spectra.

Spectrum — a colored image of a source of light.

Speed — rate of motion.

Stability — work needed to overturn a body.

Stabilizer — the part of an airplane tail which tends to keep the plane in level flight.

Standpipe — a tall reservoir used to give pressure in a water system.

States of matter — solids, liquids, and gases.

Stator — the stationary coil and frame of such machines as motors and generators, especially A C devices.

Storage cell — an electric cell which can be repeatedly charged from an electric circuit.

Strain — the deformation of a body by a force.

Stratosphere — the nearly changeless layer of the atmosphere lying above the changeable layer (troposphere).

Streamlining — giving a body a form which reduces its resistance to passing through a fluid.

Stress — force applied to deform a body.

Stringed instruments — musical instruments which produce sound by the vibration of strings.

Supercharger — an air compressor to furnish an airplane engine with sufficient oxygen at high altitudes.

Superheterodyne — a radio circuit whose action depends on beats between two sets of radio-frequency waves.

Supersonic waves — very short waves like sound waves but of a frequency above the limit of audibility for human ears. Supersonics: the science and applications of such waves.

Surface tension — the tendency of the surface of a liquid to shrink.

Suspension — a uniform distribution of an insoluble solid in a fluid.

Switch — a device for opening and closing an electric circuit.

Sympathetic vibrations — vibrations of one body produced by waves from another.

Synchronous motor — an A C motor which runs in step with the generator driving it.

Telescope — an optical instrument for forming enlarged images of distant objects. At other times it merely gathers additional light from distant objects.

Television — broadcasting scene as well as sound by radio.

Temperature — degree of hotness of a body.

Tenacity — resistance to being pulled apart.

Tetrode — a vacuum tube containing cathode, plate, grid, and screen grid.

Theodolite — a special form of telescope used by surveyors.

Thermocouple — a pair of different metals which produce an electric current when their junction is heated.

Thermograph — an instrument for recording temperature.

Thermometer — an instrument for measuring temperature.

Thermostat — a device for the automatic regulation of temperature.

Torque — moment of a force.

Transformer — a device for raising or lowering A C voltage.

Transistor — a tiny electronic device usually consisting of a germanium crystal so connected in a circuit as to perform most of the functions of the radio tubes.

Translucent bodies — objects through which some light can pass, but through which objects cannot be seen.

Transmission, automobile — the parts that connect the engine to the drive wheels, and which permit changing relative speed of engine and wheels to let the engine run at its best speed under all conditions.

Transmitter — the part of a telephone or radio set which changes sound into electrical impulses.

Transparent bodies — those that allow light from objects to pass in such a way that the objects can be distinctly seen.

Transverse waves — waves in which the vibrations are at a right angle to the wave motion.

Triode — a vacuum tube containing cathode, grid and plate, used in radio.

Tritium — a rare form of hydrogen atom having three times the mass of ordinary hydrogen.

Tropical mass — a large body of air, originating near the torrid zone.

Tuning fork — a metal fork producing a sound of definite pitch.

Turbine — a motor consisting of a wheel with curved blades, driven by water, steam, or gas.

Turbulence — eddies or disturbance produced in a fluid, generally by the passage of a solid body through it.

Ultra-violet — radiation of shorter wavelength than violet light.

Umbra — the part of a shadow from which light is entirely shut off.

Uranium — a radioactive element which may be split when struck by slow neutrons. The material of some atomic bombs.

Vacuum — in theory, a space containing no matter; in ordinary usage, a space from which most of the air or other gas has been removed, *i.e.*, *vacuum* is a relative term.

Vacuum pump — a pump for removing air or other gases from a container.

Vacuum tube — a tube, provided with electrodes, from which nearly all gas has been removed.

Vapor — the gaseous form of a substance that is a liquid (or a solid) at ordinary temperatures. (The word *gas* is commonly, but not invariably, used for substances that are gaseous at ordinary temperatures.)

Vaporization — the process of changing a liquid into a gas.

Vapor lamps — electric lamps emitting light from ionized sodium, mercury, neon, or other gases.

Vapor pressure — the pressure exerted by the vapor of an evaporating liquid.

Vector — a line used to represent the amount and direction of a force or motion.

Velocity — rate of motion in a given direction.

Velocity of light — 186,000 miles per second or 300,000 kilometers per second. A universal constant in nature.

Vibration — back-and-forth motion at a definite rate.

Video signal — the wave train into which the television camera converts the scene being televised; the receiving tube turns this back into a picture on the screen.

Virtual image — one from which light only appears to come.

Viscosity — the molecular friction that opposes the flow of a fluid.

Volt — the unit of (*a*) potential or of difference of potential; (*b*) the unit of electromotive force. In either case

$$1 \text{ volt} = \frac{1 \text{ joule of work done}}{1 \text{ coulomb of charge moved}}.$$

Voltaic cell — a device for producing an electric current by chemical action.

Voltmeter — an instrument for measuring potential, potential difference, or electromotive force.

Volume — amount of space occupied by a body.

Watt — the unit of power equivalent to 1 joule/sec. In electrical systems: watts = volts × amperes × power factor; in mechanical systems:

$$\text{watts} = \frac{\text{newtons} \times \text{meters}}{\text{seconds}}.$$

Watt-hour — unit of electrical work.

Watt-hour meter — the meter used for measuring electrical work.

Wattmeter — a device for measuring electrical power.

Waves — a set of disturbances which repeat themselves regularly and advance at a definite rate.

Wave length — distance between corresponding points in two successive waves.

Weight — the measure of the attraction between the earth and a body.

Wheatstone bridge — a special circuit used for measuring electrical resistance.

White light — light consisting of all colors in proper proportion.

Wilson cloud track chamber — an instrument for making visible the path of nuclear particles.

Wind instruments — musical instruments which produce sound by the vibration of air columns.

Wind tunnel — a structure for testing airplanes at any desired wind velocity.

Work — the accomplishment of a force acting through a distance.

X-rays — exceedingly short electromagnetic waves, produced by X-ray tubes and by radioactive substances.

Yoke — a piece of iron connecting the cores of an electromagnet.

ANSWER SHEET

TEST NO. ________________ PART ________ TITLE OF POSITION ____
(AS GIVEN IN EXAMINATION ANNOUNCEMENT - INCLUDE OPTION, IF ANY)

PLACE OF EXAMINATION ________________________________ (CITY OR TOWN) ________________ (STATE) DATE ____________

RATING

USE THE SPECIAL PENCIL. MAKE GLOSSY BLACK MARKS.

A B C D E	A B C D E	A B C D E	A B C D E	A B C D E
1	26	51	76	101
2	27	52	77	102
3	28	53	78	103
4	29	54	79	104
5	30	55	80	105
6	31	56	81	106
7	32	57	82	107
8	33	58	83	108
9	34	59	84	109
10	35	60	85	110

Make only ONE mark for each answer. Additional and stray marks may be counted as mistakes. In making corrections, erase errors COMPLETELY.

A B C D E	A B C D E	A B C D E	A B C D E	A B C D E
11	36	61	86	111
12	37	62	87	112
13	38	63	88	113
14	39	64	89	114
15	40	65	90	115
16	41	66	91	116
17	42	67	92	117
18	43	68	93	118
19	44	69	94	119
20	45	70	95	120
21	46	71	96	121
22	47	72	97	122
23	48	73	98	123
24	49	74	99	124
25	50	75	100	125

ANSWER SHEET

TEST NO. ______________ PART ______ TITLE OF POSITION ______________________________
(AS GIVEN IN EXAMINATION ANNOUNCEMENT - INCLUDE OPTION, IF ANY)

PLACE OF EXAMINATION ______________________________ (CITY OR TOWN) ______________ (STATE) DATE ______________

RATING

USE THE SPECIAL PENCIL. MAKE GLOSSY BLACK MARKS.

1 A B C D E	26 A B C D E	51 A B C D E	76 A B C D E	101 A B C D E
2 A B C D E	27 A B C D E	52 A B C D E	77 A B C D E	102 A B C D E
3 A B C D E	28 A B C D E	53 A B C D E	78 A B C D E	103 A B C D E
4 A B C D E	29 A B C D E	54 A B C D E	79 A B C D E	104 A B C D E
5 A B C D E	30 A B C D E	55 A B C D E	80 A B C D E	105 A B C D E
6 A B C D E	31 A B C D E	56 A B C D E	81 A B C D E	106 A B C D E
7 A B C D E	32 A B C D E	57 A B C D E	82 A B C D E	107 A B C D E
8 A B C D E	33 A B C D E	58 A B C D E	83 A B C D E	108 A B C D E
9 A B C D E	34 A B C D E	59 A B C D E	84 A B C D E	109 A B C D E
10 A B C D E	35 A B C D E	60 A B C D E	85 A B C D E	110 A B C D E

Make only ONE mark for each answer. Additional and stray marks may be counted as mistakes. In making corrections, erase errors COMPLETELY.

11 A B C D E	36 A B C D E	61 A B C D E	86 A B C D E	111 A B C D E
12 A B C D E	37 A B C D E	62 A B C D E	87 A B C D E	112 A B C D E
13 A B C D E	38 A B C D E	63 A B C D E	88 A B C D E	113 A B C D E
14 A B C D E	39 A B C D E	64 A B C D E	89 A B C D E	114 A B C D E
15 A B C D E	40 A B C D E	65 A B C D E	90 A B C D E	115 A B C D E
16 A B C D E	41 A B C D E	66 A B C D E	91 A B C D E	116 A B C D E
17 A B C D E	42 A B C D E	67 A B C D E	92 A B C D E	117 A B C D E
18 A B C D E	43 A B C D E	68 A B C D E	93 A B C D E	118 A B C D E
19 A B C D E	44 A B C D E	69 A B C D E	94 A B C D E	119 A B C D E
20 A B C D E	45 A B C D E	70 A B C D E	95 A B C D E	120 A B C D E
21 A B C D E	46 A B C D E	71 A B C D E	96 A B C D E	121 A B C D E
22 A B C D E	47 A B C D E	72 A B C D E	97 A B C D E	122 A B C D E
23 A B C D E	48 A B C D E	73 A B C D E	98 A B C D E	123 A B C D E
24 A B C D E	49 A B C D E	74 A B C D E	99 A B C D E	124 A B C D E
25 A B C D E	50 A B C D E	75 A B C D E	100 A B C D E	125 A B C D E